T0336423

An Introductory Course in Summability Theory

An Introductory Course in Summability Theory

Ants Aasma

Hemen Dutta

P.N. Natarajan

This edition first published 2017
© 2017 John Wiley & Sons, Inc.

The right of Ants Aasma, Hemen Dutta and P.N. Natarajan to be identified as the author(s) of this work has been asserted in accordance with law.

Registered Office
John Wiley & Sons, Inc., 111 River Street, Hoboken, NJ 07030, USA

Editorial Office
111 River Street, Hoboken, NJ 07030, USA

For details of our global editorial offices, customer services, and more information about Wiley products visit us at www.wiley.com.

Wiley also publishes its books in a variety of electronic formats and by print-on-demand. Some content that appears in standard print versions of this book may not be available in other formats.

Library of Congress Cataloguing-in-Publication Data

Names: Aasma, Ants, 1957- | Dutta, Hemen, 1981- | Natarajan, P. N., 1946
Title: An introductory course in summability theory /
    Ants Aasma, Hemen Dutta, P.N. Natarajan.
Other titles: Introductory course in summability theory
Description: Hoboken, NJ : John Wiley & Sons, Inc., 2017. | Includes
    bibliographical references and index.
Identifiers: LCCN 2017004078 (print) | LCCN 2017001841 (ebook) | ISBN 9781119397694
    (cloth) | ISBN 9781119397731 (Adobe PDF) | ISBN 9781119397779 (ePub)
Subjects: LCSH: Summability theory–Textbooks. | Sequences (Mathematics)–Textbooks.
Classification: LCC QA292 .A27 2017 (ebook) | LCC QA292 (print) | DDC 515/.243–dc23
LC record available at https://lccn.loc.gov/2017004078

Cover image: © naqiewei/Gettyimages
Cover design by Wiley

Set in 10/12pt Warnock by SPi Global, Chennai, India

Printed in the United States of America

10 9 8 7 6 5 4 3 2 1

Contents

Preface

This book is intended for graduate students and researchers as a first course in summability theory. The book is designed as a textbook as well as a reference guide for students and researchers. Any student who has a good grasp of real and complex analysis will find all the chapters within his/her reach. Knowledge of functional analysis will be an added asset. Several problems are also included in the chapters as solved examples and chapter-end exercises along with hints, wherever felt necessary. The book consists of nine chapters and is organized as follows:

In chapter 1, after a very brief introduction to summability theory, general matrix methods are introduced and Silverman–Toeplitz theorem on regular matrices is proved. Schur's, Hahn's, and Knopp–Lorentz theorems are then taken up. Steinhaus theorem that a matrix cannot be both regular and a Schur matrix is then deduced.

Chapter 2 is devoted to a study of some special summability methods. The Nörlund method, the Weighted Mean method, the Abel method, and the $(C; 1)$ method are introduced, and their properties are discussed.

Chapter 3 is devoted to a study of some more special summability methods. The $(M; \lambda_n)$ method, the Euler method, the Borel method, the Taylor method, the Hölder and Cesàro methods, and the Hausdorff method are introduced, and their properties are discussed.

In Chapter 4, various Tauberian theorems involving certain summability methods are discussed.

In Chapter 5, matrix transforms of summability and absolute summability domains of reversible and normal methods are studied. The notion of M-consistency of matrix methods A and B is introduced and its properties are studied. As a special case, some inclusion problems are analyzed.

In Chapter 6, the notion of a perfect matrix method is introduced. Matrix transforms of summability domains of regular perfect matrix methods are considered.

In Chapter 7, matrix transforms of summability and absolute summability domains of the Cesàro and the Riesz methods are studied. Also, some special classes of matrices transforming the summability or absolute summability domain of a matrix method into the summability or absolute summability domain of another matrix method are considered.

In Chapter 8, the notions of the convergence and the boundedness of sequences with speed λ (λ is a positive monotonically increasing sequence) are introduced. The necessary and sufficient conditions for a matrix A to be transformed from the set of all λ-bounded or λ-convergent sequences into the set of all λ-bounded or μ-convergent sequences (μ is another speed) are described. In addition, the notions of the summability and the boundedness with speed by a matrix method are introduced and their properties are described. Also, the M-consistency of matrix methods A and B on the set of all sequences, λ-bounded by A, is investigated. As applications of main results, the matrix transforms for the case of Riesz methods are investigated, and the comparison of approximation orders of Fourier expansions in Banach spaces by different matrix methods is studied.

Chapter 9 continues the investigation of convergence, boundedness, and summability with speed, started in Chapter 8. Some topological properties of the spaces m^λ (the set of all λ bounded sequences), c^λ (the set of all λ-convergent sequences), c^λ_A (the set of all sequences, λ- convergent by A), and m^λ_A (the set of all sequences, λ-bounded by A) are introduced. The notions of λ-reversible, λ-perfect, and λ-conservative matrix methods are introduced. The necessary and sufficient conditions for a matrix M to be transformed from c^λ_A into c^λ_B or into m^μ_B are described. Also, the M-consistency of matrix methods A and B on c^λ_A is investigated. As applications of main results, the matrix transforms for the cases of Riesz and Cesàro methods are investigated.

We were influenced by the work of several authors during the preparation of the text. Constructive criticism, comments, and suggestions for the improvement of the contents of the book are always welcome. The authors are thankful to several researchers and colleagues for their valuable suggestions. Special thanks to Billy E. Rhoades, emeritus professor, Indiana University, USA, for editing the final draft of the book.

Ants Aasma, Tallinn, Estonia
Hemen Dutta, Guwahati, India
P.N. Natarajan, Chennai, India
December, 2016

About the Authors

Ants Aasma is an associate professor of mathematical economics in the department of economics and finance at Tallinn University of Technology, Estonia. He received his PhD in mathematics in 1993 from Tartu University, Estonia. His main research interests include topics from the summability theory, such as matrix methods, matrix transforms, summability with speed, convergence acceleration, and statistical convergence. He has published several papers on these topics in reputable journals and visited several foreign institutions in connection with conferences. Dr. Aasma is also interested in approximation theory and dynamical systems in economics. He is a reviewer for several journals and databases of mathematics. He is a member of some mathematical societies, such as the Estonian Mathematical Society and the Estonian Operational Research Society. He teaches real analysis, complex analysis, operations research, mathematical economics, and financial mathematics. Dr. Aasma is the author of several textbooks for Estonian universities.

Hemen Dutta is a senior assistant professor of mathematics at Gauhati University, India. Dr. Dutta received his MSc and PhD in mathematics from Gauhati University, India. He received his MPhil in mathematics from Madurai Kamaraj University, India. Dr. Dutta's research interests include summability theory and functional analysis. He has to his credit several papers in research journals and two books. He visited foreign institutions in connection with research collaboration and conference. He has delivered talks at foreign and national institutions. He is a member on the editorial board of several journals and he is continuously reviewing for some databases and journals of mathematics. Dr. Dutta is a member of some mathematical societies.

P.N. Natarajan, Dr Radhakrishnan Awardee for the Best Teacher in Mathematics for the year 1990–91 by the Government of Tamil Nadu, India, has been working as an independent researcher after his retirement, in 2004, as professor and head, department of mathematics, Ramakrishna Mission Vivekananda College, Chennai, Tamil Nadu, India. Dr. Natarajan received his PhD in analysis from the University of Madras in 1980. He has to his credit over 100 research papers published in several reputed international journals. He authored a book (two editions) and contributed in an edited book. Dr. Natarajan's research interests include summability theory and functional analysis (both classical and ultrametric). Besides visiting several institutes of repute in Canada, France, Holland, and Greece on invitation, he has participated in several international conferences and has chaired sessions.

About the Book

This book is designed as a textbook for graduate students and researchers as a first course in summability theory. The book starts with a short and compact overview of basic results on summability theory and special summability methods. Then, results on matrix transforms of several matrix methods are discussed, which have not been widely discussed in textbooks yet. One of the most important applications of summability theory is the estimation of the speed of convergence of a sequence or series. In the textbooks published in English language until now, no description of the notions of convergence, boundedness, and summability with speed can be found, started by G. Kangro in 1969. Finally, this book discusses these concepts and some applications of these concepts in approximation theory. Each chapter of the book contains several solved examples and chapter-end exercises including hints for solution.

1

Introduction and General Matrix Methods

1.1 Brief Introduction

The study of the convergence of infinite series is an ancient art. In ancient times, people were more concerned with orthodox examinations of convergence of infinite series. Series that did not converge were of no interest to them until the advent of L. Euler (1707–1783), who took up a serious study of "divergent series"; that is, series that did not converge. Euler was followed by a galaxy of great mathematicians, such as C.F. Gauss (1777–1855), A.L. Cauchy (1789–1857), and N.H. Abel (1802–1829). The interest in the study of divergent series temporarily declined in the second half of the nineteenth century. It was rekindled at a later date by E. Cesàro, who introduced the idea of $(C, 1)$ convergence in 1890. Since then, many other mathematicians have been contributing to the study of divergent series. Divergent series have been the motivating factor for the introduction of summability theory.

Summability theory has many uses in analysis and applied mathematics. An engineer or physicist who works with Fourier series, Fourier transforms, or analytic continuation can find summability theory very useful for his/her research.

Throughout this chapter, we assume that all indices and summation indices run from 0 to ∞, unless otherwise specified. We denote sequences by $\{x_k\}$ or (x_k), depending on convenience.

Consider the sequence

$$\{s_n\} = \{1, 0, 1, 0, \dots\},$$

which is known to diverge. However, let

$$t_n = \frac{s_0 + s_1 + \cdots + s_n}{n + 1},$$

i.e.,
$$t_n = \begin{cases} \frac{k+1}{2k+1}, & \text{if } n = 2k; \\ \frac{k+1}{2k+2}, & \text{if } n = 2k + 1, \end{cases}$$

An Introductory Course in Summability Theory, First Edition. Ants Aasma, Hemen Dutta, and P.N. Natarajan.
© 2017 John Wiley & Sons, Inc. Published 2017 by John Wiley & Sons, Inc.

proving that

$$t_n \to 1/2, \; n \to \infty.$$

In this case, we say that the sequence $\{s_n\}$ converges to $1/2$ in the sense of Cesàro or $\{s_n\}$ is $(C, 1)$ summable to $1/2$. Similarly, consider the infinite series

$$\sum_n a_n = 1 - 1 + 1 - 1 + \cdots.$$

The associated sequence $\{s_n\}$ of partial sums is $\{1, 0, 1, 0, \dots\}$, which is $(C, 1)$-summable to $1/2$. In this case, we say that the series $\sum_n a_n = 1 - 1 + 1 - 1 + \cdots$ is $(C, 1)$-summable to $1/2$.

With this brief introduction, we recall the following concepts and results.

1.2 General Matrix Methods

Definition 1.1 Given an infinite matrix $A = (a_{nk})$, and a sequence $x = \{x_k\}$, by the A-transform of $x = \{x_k\}$, we mean the sequence

$$A(x) = \{(Ax)_n\},$$
$$(Ax)_n = \Sigma_k a_{nk} x_k,$$

where we suppose that the series on the right converges. If $\lim_{n \to \infty} (Ax)_n = t$, we say that the sequence $x = \{x_k\}$ is summable A or A-summable to t. If $\lim_{n \to \infty} (Ax)_n = t$ whenever $\lim_{k \to \infty} x_k = s$, then A is said to be preserving convergence for convergent sequences, or sequence-to-sequence conservative (for brevity, Sq-Sq conservative). If A is sequence-to-sequence conservative with $s = t$, we say that A is sequence-to-sequence regular (shortly, Sq-Sq regular). If $\lim_{n \to \infty} (Ax)_n = t$, whenever, $\Sigma_k x_k = s$, then A is said to preserve the convergence of series, or series-to-sequence conservative (i.e., Sr-Sq conservative). If A is series-to-sequence conservative with $s = t$, we say that A is series-to-sequence regular (shortly, Sr-Sq regular).

In this chapter and in Chapters 2 and 3, for conservative and regular, we mean only Sq–Sq conservativity and Sq-Sq regularity.

If X, Y are sequence spaces, we write

$$A \in (X, Y),$$

if $\{(Ax)_n\}$ is defined and $\{(Ax)_n\} \in Y$, whenever, $x = \{x_k\} \in X$. With this notation, if A is conservative, we can write $A \in (c, c)$, where c denotes the set of all convergent sequences. If A is regular, we write

$$A \in (c, c; P),$$

P denoting the "preservation of limit."

Definition 1.2 A method $A = (a_{nk})$ is said to be lower triangular (or simply, triangular) if $a_{nk} = 0$ for $k > n$, and normal if A is lower triangular if $a_{nn} \neq 0$ for every n.

Example 1.1 Let A be the Zweier method; that is, $A = Z_{1/2}$, defined by the lower triangular method $A = (a_{nk})$ where (see [2], p. 14) $a_{00} = 1/2$ and

$$a_{nk} = \begin{cases} \frac{1}{2}, & \text{if } k = n - 1 \text{ and } k = n; \\ 0, & \text{if } k < n - 1 \end{cases}$$

for $n \geq 1$. The method $A = Z_{1/2}$ is regular. The transformation $(Ax)_n$ for $n \geq 1$ can be presented as

$$(Ax)_n = \frac{x_{n-1} + x_n}{2}.$$

Then,

$$\lim_n (Ax)_n = \lim_k x_k$$

for every $x = \{x_k\} \in c$; that is, $Z_{1/2} \in (c, c; P)$.

We now prove a landmark theorem in summability theory due to Silverman–Toeplitz, which characterizes a regular matrix in terms of the entries of the matrix (see [3–5]).

Theorem 1.1 (Silverman-Toeplitz) $A = (a_{nk})$ is regular, that is, $A \in (c, c; P)$, if and only if

$$\sup_{n \geq 0} \sum_k |a_{nk}| < \infty; \tag{1.1}$$

$$\lim_{n \to \infty} a_{nk} := \delta_k; \tag{1.2}$$

and

$$\lim_{n \to \infty} \sum_k a_{nk} = \delta \tag{1.3}$$

with $\delta_k \equiv 0$ and $\delta \equiv 1$.

Proof: Sufficiency. Assume that conditions (1.1)–(1.3) with $\delta_k \equiv 0$ and $\delta \equiv 1$ hold. Let $x = \{x_k\} \in c$ with $\lim_{k \to \infty} x_k = s$. Since $\{x_k\}$ converges, it is bounded; that is, $x_k = O(1)$, $k \to \infty$, or, equivalently, $|x_k| \leq M$, $M > 0$ for all k.

Now

$$\sum_k |a_{nk} x_k| \leq M \sum_k |a_{nk}| < \infty,$$

in view of (1.1), and so

$$(Ax)_n = \sum_k a_{nk} x_k$$

is defined. Now

$$(Ax)_n = \sum_k a_{nk}(x_k - s) + s \sum_k a_{nk}. \tag{1.4}$$

Since $\lim_{k\to\infty} x_k = s$, given an $\epsilon > 0$, there exists an $n \in \mathbb{N}$, where $\mathbb{N}$ denotes the set of all positive integers, such that

$$|x_k - s| < \frac{\epsilon}{2L}, \quad k > N, \tag{1.5}$$

where $L > 0$ is such that

$$|x_n - s| \leq L, \quad \sum_k |a_{nk}| \leq L, \tag{1.6}$$

and hence

$$\sum_k a_{nk}(x_k - s) = \sum_{k=0}^{N} a_{nk}(x_k - s) + \sum_{k=N+1}^{\infty} a_{nk}(x_k - s),$$

$$\left| \sum_k a_{nk}(x_k - s) \right| \leq \sum_{k=0}^{N} |a_{nk}||x_k - s| + \sum_{k=N+1}^{\infty} |a_{nk}||x_k - s|.$$

Using (1.5) and (1.6), we obtain

$$\sum_{k=N+1}^{\infty} |a_{nk}||x_k - s| \leq \frac{\epsilon}{2L} \sum_k |a_{nk}| \leq \frac{\epsilon}{2L} L = \frac{\epsilon}{2}.$$

By (1.2), there exists a positive integer n_0 such that

$$|a_{nk}| < \frac{\epsilon}{2L(N+1)}, \quad k = 0, 1, \dots, N, \text{ for } n > n_0.$$

This implies that

$$\sum_{k=0}^{N} |a_{nk}||x_k - s| < L(N+1) \frac{\epsilon}{2L(N+1)} = \frac{\epsilon}{2}, \quad \text{for } n > n_0.$$

Consequently, for every $\epsilon > 0$, we have

$$\left| \sum_k a_{nk}(x_k - s) \right| < \frac{\epsilon}{2} + \frac{\epsilon}{2} = \epsilon \text{ for } n > n_0.$$

Thus,

$$\lim_{n\to\infty} \sum_k a_{nk}(x_k - s) = 0. \tag{1.7}$$

Taking the limit as $n \to \infty$ in (1.4), we have, by (1.7), that

$$\lim_{n\to\infty} (Ax)_n = s,$$

since $\delta = 1$. Hence, A is regular, completing the proof of the sufficiency part.

Necessity. Let A be regular. For every fixed k, consider the sequence $x = \{x_n\}$, where

$$
x_n = \begin{cases} 1, & n = k; \\ 0, & \text{otherwise.} \end{cases}
$$

For this sequence x, $(Ax)_n = a_{nk}$. Since $\lim_{n \to \infty} x_n = 0$ and A is regular, it follows that $\delta_k \equiv 0$. Again consider the sequence $x = \{x_n\}$, where $x_n = 1$ for all n. Note that $\lim_{n \to \infty} x_n = 1$. For this sequence x, $(Ax)_n = \Sigma_k a_{nk}$. Since $\lim_{n \to \infty} x_n = 1$ and A is regular, we have $\delta = 1$. It remains to prove (1.1). First, we prove that $\Sigma_k |a_{nk}|$ converges. Suppose not. Then, there exists an $N \in \mathbb{N}$ such that

$$
\sum_k |a_{Nk}| \text{ diverges.}
$$

In fact, $\Sigma_k |a_{Nk}|$ diverges to ∞. So we can find a strictly increasing sequence $k(j)$ of positive integers such that

$$
\sum_{k=k(j-1)}^{k(j)-1} |a_{Nk}| > 1, \quad j = 1, 2, \dots . \tag{1.8}
$$

Define the sequence $x = \{x_k\}$ by

$$
x_k = \begin{cases} \dfrac{|a_{Nk}|}{j a_{Nk}}, & \text{if } a_{Nk} \neq 0 \text{ and } k(j-1) \leq k < k(j), \ j = 1, 2, \dots ; \\ 0, & \text{if } k = 0 \text{ or } a_{Nk} = 0. \end{cases}
$$

Note that $\lim_{k \to \infty} x_k = 0$ and $\Sigma_k a_{nk} x_k$ converges. In particular, $\Sigma_k a_{Nk} x_k$ converges. However,

$$
\sum_k a_{Nk} x_k = \sum_{j=1}^{\infty} \sum_{k=k(j-1)}^{k(j)-1} \frac{|a_{Nk}|}{j} = \sum_{j=1}^{\infty} \frac{1}{j} \sum_{k=k(j-1)}^{k(j)-1} |a_{Nk}| > \sum_{j=1}^{\infty} \frac{1}{j}.
$$

This leads to a contradiction since $\sum_{j=1}^{\infty} \frac{1}{j}$ diverges. Thus,

$$
\sum_k |a_{nk}| \text{ converges for every } n \in \mathbb{N}.
$$

To prove that (1.1) holds, we assume that

$$
\sup_{n \geq 0} \sum_k |a_{nk}| = \infty
$$

and arrive at a contradiction.

We construct two strictly increasing sequences $\{m(j)\}$ and $\{n(j)\}$ of positive integers in the following manner.

Let $m(0) = 0$. Since $\Sigma_k |a_{m(0),k}| < \infty$, choose $n(0)$ such that

$$\sum_{k=n(0)+1}^{\infty} |a_{m(0),k}| < 1.$$

Having chosen the positive integers $m(0), m(1), \ldots, m(j-1)$ and $n(0), n(1), \ldots,$ $n(j-1)$, choose positive integers $m(j) > m(j-1)$ and $n(j) > n(j-1)$ such that

$$\sum_{k} |a_{m(j),k}| > j^2 + 2j + 2; \qquad (1.9)$$

$$\sum_{k=0}^{n(j-1)} |a_{m(j),k}| < 1; \qquad (1.10)$$

and

$$\sum_{k=n(j)+1}^{\infty} |a_{m(j),k}| < 1. \qquad (1.11)$$

Now define the sequence $x = \{x_k\}$, where

$$x_k = \begin{cases} \dfrac{|a_{m(j),k}|}{ja_{m(j),k}}, & \text{if } n(j-1) < k \leq n(j), a_{m(j),k} \neq 0, j = 1, 2, \ldots; \\ 0, & \text{otherwise.} \end{cases}$$

Note that $\lim_{k \to \infty} x_k = 0$. Since A is regular, $\lim_{n \to \infty} (Ax)_n = 0$. However, using (1.9)–(1.11), we have

$$|(Ax)_{m(j)}| = \left| \sum_k a_{m(j),k} x_k \right|$$

$$= \left| \sum_{k=0}^{n(j-1)} a_{m(j),k} x_k + \sum_{k=n(j-1)+1}^{n(j)} a_{m(j),k} x_k + \sum_{k=n(j)+1}^{\infty} a_{m(j),k} x_k \right|$$

$$\geq \left| \sum_{k=n(j-1)+1}^{n(j)} a_{m(j),k} x_k \right| - \sum_{k=0}^{n(j-1)} |a_{m(j),k} x_k| - \sum_{k=n(j)+1}^{\infty} |a_{m(j),k} x_k|$$

$$> \frac{1}{j} \sum_{k=n(j-1)+1}^{n(j)} |a_{m(j),k}| - 1 - 1$$

$$= \frac{1}{j} \left[\sum_{k} |a_{m(j),k}| - \sum_{k=0}^{n(j-1)} |a_{m(j),k}| - \sum_{k=n(j)+1}^{\infty} |a_{m(j),k}| \right] - 2$$

$$> \frac{1}{j}[(j^2 + 2j + 2) - 1 - 1] - 2 = j + 2 - 2 = j, j = 1, 2, \ldots.$$

Thus, $\{(Ax)_{m(j)}\}$ diverges, which contradicts the fact that $\{(Ax)_n\}$ converges. Consequently, (1.1) holds. This completes the proof of the theorem.

Example 1.2 Let A be the Cesàro method $(C, 1)$; that is, $A = (C, 1)$. This method is defined by the lower triangular matrix $A = (a_{nk})$, where $a_{nk} = 1/(n+1)$ for all $k \leq n$. It is easy to see that all of the conditions of Theorem 1 are satisfied. Hence, $(C, 1) \in (c, c; P)$.

Example 1.3 Let $A_{-1,1}$ be the method defined by the lower triangular matrix (a_{nk}), where $a_{00} = 1$ and

$$a_{nk} = \begin{cases} -1, & \text{if } k = n - 1; \\ 1, & \text{if } k = n; \\ 0, & \text{if } k < n - 1 \end{cases}$$

for $n \geq 1$. It is easy to see that, in this case, $\delta_k \equiv 0$, $\delta = 0 \neq 1$ and condition (1.1) holds. Therefore, $A_{-1,1}$ does not belong to $(c, c; P)$. However, $A_{-1,1} \in (c, c)$ and $A_{-1,1} \in (c_0, c_0)$, where c_0 denotes the set of all sequences converging to 0 (see Exercises 1.1 and 1.4).

Let m (or ℓ_∞) denote the set of all bounded sequences. For $x = \{x_k\} \in \ell_\infty$, define

$$\| x \| = \sup_{k \geq 0} |x_k|. \tag{1.12}$$

Then, it is easy to see that m is a Banach space and c is a closed subspace of m with respect to the norm defined by (1.12).

Definition 1.3 The matrix $A = (a_{nk})$ is called a Schur matrix if $A \in (m, c)$; that is, $\{(Ax)_n\} \in c$, whenever, $x = \{x_k\} \in m$.

The following result gives a characterization of a Schur matrix in terms of the entries of the matrix (see [3–5]).

Theorem 1.2 (Schur) $A = (a_{nk})$ is a Schur matrix if and only if (1.2) holds and

$$\sum_k |a_{nk}| \text{ converges uniformly in } n. \tag{1.13}$$

Proof: Sufficiency. Assume that (1.2) and (1.13) hold. Then, (1.13) implies that the series $\sum_k |a_{nk}|$ converge, n belongs to N. By (1.2) and (1.13), we obtain that

$$\sup_{n \geq 0} \sum_k |a_{nk}| = M < \infty.$$

Thus, for each r, we have

$$\lim_{n \to \infty} \sum_{k=0}^{r} |a_{nk}| \leq M.$$

Hence,

$$\sum_{k=0}^{r} |\delta_k| \leq M \text{ for every } r,$$

and so

$$\sum_{k} |\delta_k| < \infty.$$

Thus, if $x = \{x_k\} \in m$, it follows that $\Sigma_k a_{nk} x_k$ converges absolutely and uniformly in n. Consequently,

$$\lim_{n \to \infty} (Ax)_n = \lim_{n \to \infty} \sum_{k} a_{nk} x_k = \sum_{k} \delta_k x_k,$$

proving that $\{(Ax)_n\} \in c$; that is, $A \in (m, c)$, proving the sufficiency part.

Necessity. Let $A = (a_{nk}) \in (m, c)$. Then, $A \in (c, c)$ and so (1.2) holds. Again, since $A \in (c, c)$, we get that (1.1) holds; that is,

$$\sup_{n \geq 0} \sum_{k} |a_{nk}| < \infty.$$

As in the sufficiency part of the present theorem, it follows that $\Sigma_k |\delta_k| < \infty$. We write

$$b_{nk} = a_{nk} - \delta_k.$$

Then, $\{\Sigma_k b_{nk} x_k\}$ converges for all $x = \{x_k\} \in m$. We now claim that

$$\sum_{k} |b_{nk}| \to 0, n \to \infty. \tag{1.14}$$

Suppose not. Then,

$$\overline{\lim_{n \to \infty}} \sum_{k} |b_{nk}| = c > 0.$$

So,

$$\sum_{k} |b_{mk}| \to c, m \to \infty$$

through some subsequence of positive integers. We also note that

$$\lim_{m \to \infty} b_{mk} = 0 \text{ for all } k \in \mathbb{N}.$$

We can now find a positive integer $m(1)$ such that

$$\left| \sum_{k} |b_{m(1),k}| - c \right| < \frac{c}{10}$$

and

$$|b_{m(1),0}| + |b_{m(1),1}| < \frac{c}{10}.$$

Since $\Sigma_k |b_{m(1),k}| < \infty$, we can choose $k(2) > 1$ such that

$$\sum_{k=k(2)+1}^{\infty} |b_{m(1),k}| < \frac{c}{10}.$$

It now follows that

$$\left| \sum_{k=2}^{k(2)} |b_{m(1),k}| - c \right| = \left| \left(\sum_{k} |b_{m(1),k}| - c \right) - (|b_{m(1),0}| + |b_{m(1),1}|) - \sum_{k=k(2)+1}^{\infty} |b_{m(1),k}| \right|$$

$$< \frac{c}{10} + \frac{c}{10} + \frac{c}{10} = \frac{3c}{10}.$$

Now choose a positive integer $m(2) > m(1)$ such that

$$\left| \sum_{k} |b_{m(2),k}| - c \right| < \frac{c}{10}$$

and

$$\sum_{k=0}^{k(2)} |b_{m(2),k}| < \frac{c}{10}.$$

Then, choose a positive integer $k(3) > k(2)$ such that

$$\sum_{k=k(3)+1}^{\infty} |b_{m(2),k}| < \frac{c}{10}.$$

It now follows that

$$\left| \sum_{k=k(2)+1}^{k(3)} |b_{m(2),k}| - c \right| < \frac{3c}{10}.$$

Continuing this way, we find $m(1) < m(2) < \cdots$ and $1 = k(1) < k(2) < k(3) < \cdots$ so that

$$\sum_{k=0}^{k(r)} |b_{m(r),k}| < \frac{c}{10}; \tag{1.15}$$

$$\sum_{k=k(r+1)+1}^{\infty} |b_{m(r),k}| < \frac{c}{10}; \tag{1.16}$$

and

$$\left| \sum_{k=k(r)+1}^{k(r+1)} |b_{m(r),k}| - c \right| < \frac{3c}{10}. \tag{1.17}$$

We now define a sequence $x = \{x_k\}$ as follows: $x_0 = x_1 = 0$ and

$$x_k = (-1)^r \operatorname{sgn} b_{m(r),k},$$

if $k(r) < k \le k(r+1)$, $r = 1, 2, \ldots$. Note that $x = \{x_k\} \in m$ and $\|x\| = 1$. Now

$$
\left| \sum_k b_{m(r),k} x_k - (-1)^r c \right| = \left| \sum_{k=0}^{k(r)} b_{m(r),k} x_k + \sum_{k=k(r)+1}^{k(r+1)} b_{m(r),k} x_k \right.
$$

$$
\left. + \sum_{k=k(r+1)+1}^{\infty} b_{m(r),k} x_k - (-1)^r c \right|
$$

$$
= \left| \left\{ \sum_{k=k(r)+1}^{k(r+1)} |b_{m(r),k}| - c \right\} (-1)^r \right.
$$

$$
\left. + \sum_{k=0}^{k(r)} b_{m(r),k} x_k + \sum_{k=k(r+1)+1}^{\infty} b_{m(r),k} x_k \right|
$$

$$
< \frac{3c}{10} + \frac{c}{10} + \frac{c}{10} = \frac{c}{2},
$$

using (1.15), (1.16) and (1.17).

Consequently, $\{\sum_k b_{nk} x_k\}$ is not a Cauchy sequence and so it is not convergent, which is a contradiction. Thus, (1.14) holds. So, given $\epsilon > 0$, there exists a positive integer n_0 such that

$$
\sum_k |b_{nk}| < \epsilon, \quad n > n_0. \tag{1.18}
$$

Since $\sum_k |b_{nk}| < \infty$ for $0 \le n \le n_0$, we can find a positive integer M such that

$$
\sum_{k=M}^{\infty} |b_{nk}| < \epsilon, \quad 0 \le n \le n_0. \tag{1.19}
$$

In view of (1.18) and (1.19), we have

$$
\sum_{k=M}^{\infty} |b_{nk}| < \epsilon \quad \text{for all } n,
$$

that is, $\sum_k |b_{nk}|$ converges uniformly in n. Since $\sum_k |\delta_k| < \infty$, it follows that $\sum_k |a_{nk}|$ converges uniformly in n, proving the necessity part. The proof of the theorem is now complete.

Example 1.4 Let $A = (a_{nk})$ be defined by the lower triangular matrix

$$
a_{nk} := \frac{1}{(n+1)(k+1)}. \tag{1.20}
$$

Then, $\delta_k = 0$ and

$$
\sum_k |a_{nk}| = \frac{1}{n+1} \sum_{k=0}^{n} \frac{1}{k+1} = \frac{1}{n+1} O(\ln(n+1)) \to 0 \quad \text{if } n \to \infty;
$$

that is, condition (1.13) is fulfilled. Hence, $A \in (m, c)$ by Theorem 1.2.

Using Theorems 1.1 and 1.2, we can deduce the following important result.

Theorem 1.3 (Steinhaus) An infinite matrix cannot be both regular and a Schur matrix. In other words, given a regular matrix, there exists a bounded, divergent sequence which is not A-summable.

Proof: Let A be a regular and a Schur matrix. Then, (1.2) and (1.3) hold with $\delta_k \equiv 0$ and $\delta \equiv 1$. Using (1.13), we get

$$\lim_{n\to\infty} \sum_k a_{nk} = \sum_k \left(\lim_{n\to\infty} a_{nk} \right) = 0$$

by (1.2), which contradicts (1.3). This establishes our claim.

For the proof of the following results, we need some additional notations. Let

$$cs := \left\{ x = (x_k) \ : \ (X_n) \in c;\ X_n := \sum_{k=0}^{n} x_k \right\},$$

$$cs_0 := \left\{ x = (x_k) \mid (X_n) \in c_0;\ X_n := \sum_{k=0}^{n} x_k \right\},$$

$$l := \left\{ x = (x_k) \ : \ \sum_k |x_k| < \infty \right\},$$

$$bv := \{ x = (x_k) \ : \ (\Delta x_k) \in l \},$$

where

$$\Delta x_k := (\Delta^1 x)_k = x_k - x_{k+1},$$

and

$$bv_0 := bv \cap c_0.$$

It is easy to see that the set of sequences cs is equivalent to the set of all convergent series.

Theorem 1.4 (Hahn) Let $A = (a_{nk})$ be a matrix method. Then, $A \in (l, c)$ if and only if condition (1.2) holds and

$$a_{nk} = O(1). \tag{1.21}$$

Proof: For every fixed k, let e^k be the sequence in which 1 occurs in the $(k + 1)^{\text{th}}$ place and 0 elsewhere. As $e^k \in l$, then condition (1.2) is necessary. It is easy to see that we can consider a matrix A as a continuous linear operator from l to c with the norm $\|A\| = \sup_{n,k} |a_{nk}|$. The proof now follows from the Banach–Steinhaus theorem.

Example 1.5 It is easy to see that the methods $Z_{1/2}$, $(C, 1)$, $A_{-1,1}$ and the methods A, defined by (1.20), considered in Examples 1.1–1.4, belong (l, c).

Theorem 1.5 Let $A = (a_{nk})$ be a matrix method. Then, $A \in (cs, c)$ if and only if condition (1.2) holds and

$$\sum_k |\Delta_k a_{nk}| = O(1).$$ (1.22)

Moreover,

$$\lim_n A_n x = \delta_0 \lim Sx + \sum_k \Delta \delta_k (X_k - \lim Sx)$$ (1.23)

for every $x := (x_k) \in cs$.

Proof: First, we find conditions for the existence of the transform Ax for every $x \in cs$. Define

$$y_n^m = \sum_{k=0}^m a_{nk} x_k$$ (1.24)

for $x := (x_k) \in cs$. Using the Abel's transform (see, e.g., [1], p. 18)

$$\sum_{k=0}^m \epsilon_k x_k = \sum_{k=0}^{m-1} \Delta \epsilon_k X_k + \epsilon_m X_m,$$

where

$$X_k := \sum_{l=0}^k x_l,$$ (1.25)

we can write

$$y_n^m = \sum_{k=0}^{m-1} \Delta a_{nk} X_k + a_{nm} X_m.$$

This implies that Ax exists for every $x \in cs$ if and only

$$y_n := \lim_m y_n^m$$ (1.26)

has a finite limit for every $(X_k) \in c$ and $n \in \mathbb{N}$, since, for every $(X_k) \in c$ there exists an $(x_l) \in cs$, such that (1.25) holds. Hence, for every $n \in \mathbb{N}$, the limit y_n in (1.26) exists for every $(X_k) \in c$ if and only if the matrix $D^n := (d_{mk}^n) \in (c, c)$, where

$$d_{mk}^n = \begin{cases} \Delta_k a_{nk}, & \text{if } k < m; \\ a_{nm}, & \text{if } k = m; \\ 0, & \text{if } k > m. \end{cases}$$

As

$$\lim_m d^n_{mk} = \Delta_k a_{nk} \text{ and } \sum_{k=0}^{m} d^n_{mk} = a_{n0},$$

$D^n \in (c, c)$ if and only if

$$\sum_{k=0}^{m-1} |\Delta_k a_{nk}| + |a_{nm}| = O_n(1). \tag{1.27}$$

As

$$|a_{nm}| \leq \sum_{k=0}^{m-1} |\Delta_k a_{nk}| + |a_{n0}|,$$

(1.27) is equivalent to the condition

$$\sum_k |\Delta_k a_{nk}| = O_n(1). \tag{1.28}$$

Moreover, using (1.38) (see Exercise 1.1), from the existence of the limits in (1.26) we obtain

$$y_n - a_{n0} \lim_k X_k = \sum_k (\Delta_k a_{nk})(X_k - \lim_k X_k). \tag{1.29}$$

As $(X_k - \lim_k X_k) \in c_0$, then, using Exercise 1.3, we conclude that transform (1.29) and the finite limit $\lim_n (y_n - a_{n0} \lim_k X_k)$ exists if and only if

there exists the finite limit $\lim_n \Delta_k a_{nk} := d_k$ (1.30)

and condition (1.22) holds. We note also that condition (1.28) follows from (1.22). In addition,

$$\lim_n (y_n - a_{n0} \lim_k X_k) = \sum_k d_k (X_k - \lim_k X_k). \tag{1.31}$$

Now (1.31) implies that, for the existence of the finite limit $\lim_n y_n$ it is necessary that

there exists the finite limit δ_0. (1.32)

Therefore, using (1.30), we obtain that, for every $k \geq 1$ the existence of finite limits δ_k; that is, condition (1.2) is necessary. From the other side, condition (1.30) follows from (1.2).

Finally, the validity of (1.23) follows from (1.31) and (1.32).

Example 1.6 Let A be the Zygmund method of order 1; that is, $A = Z^1$, defined by the lower triangular matrix (a_{nk}), where

$$a_{nk} = 1 - \frac{k}{n+1}.$$

Then, $\delta_k \equiv 1$ and $\Delta a_{nk} = 1/(n+1)$ for every $k \le n$. Hence,

$$\sum_k |\Delta_k a_{nk}| = 1;$$

that is, condition (1.22) holds. Thus, $Z^1 \in (cs, c)$ by Theorem 1.5. Moreover, Z^1 is Sr–Sq regular (see Exercise 1.5).

Theorem 1.6   Let $A = (a_{nk})$ be a matrix method. Then, $A \in (bv, c)$ if and only if conditions (1.2), (1.3) hold and

$$\sum_{k=0}^{m} a_{nk} = O(1). \tag{1.33}$$

Moreover, for (bv_0, c) condition (1.3) is redundant.

Proof: As $e \in bv$, then it is necessary for $A \in (bv, c)$ that

$$\text{all series } \sum_{k=l}^{\infty} a_{nk} \text{ converge.} \tag{1.34}$$

Hence,

$$\sum_{k=0}^{m} a_{nk} = O_n(1).$$

Let

$$x_k - \lim x_k := v_k$$

and y_n^m be defined by (1.24) for every $x = (x_k) \in bv$. As $(v_k) \in c_0$ and

$$\sum_k |v_k - v_{k-1}| < \infty \quad (v_{-1} = 0)$$

(i.e., $(v_k - v_{k-1}) \in l$), then

$$y_n^m - \lim_k x_k \sum_{k=0}^{m} a_{nk} = \sum_{k=0}^{m} a_{nk} v_k = - \sum_{k=0}^{m} a_{nk} \sum_{l=k+1}^{\infty} (v_l - v_{l-1})$$

$$= - \sum_{l=1}^{m} \left(\sum_{k=0}^{l-1} a_{nk} \right) (v_l - v_{l-1}) - \sum_{l=m+1}^{\infty} \left(\sum_{k=0}^{m} a_{nk} \right) (v_l - v_{l-1}).$$

Hence, for $m \to \infty$, we obtain

$$y_n - \lim_k x_k \sum_k a_{nk} = - \sum_{l=1}^{\infty} \left(\sum_{k=0}^{l-1} a_{nk} \right) (v_l - v_{l-1}), \tag{1.35}$$

where $\lim_m y_n^m = y_n$. Thus, transformation (1.35) exists if condition (1.34) holds. So we can conclude from (1.35) that conditions (1.2) and (1.33) are necessary and sufficient for the existence of the finite limit $\lim_n \left(y_n - \lim_k x_k \sum_k a_{nk} \right)$ by Theorem 1.4, since the existence of the finite limits

$$\lim_n \sum_{k=0}^{l-1} a_{nk}$$

is equivalent to (1.2). As $e \in bv$, then condition (1.3) is necessary. Therefore, from the existence of the finite limit $\lim_n \left(y_n - \lim_k x_k \sum_k a_{nk} \right)$ follows $(y_n) \in c$ for every $x \in bv$.

It is easy to see that for (bv_0, c) the existence of the finite limit δ is redundant.

Example 1.7 As $bv \subset c \subset m$, then (see Examples 1.1–1.4) the methods $Z_{1/2}$, $(C, 1)$, $A_{-1,1}$ and the method A, defined by (1.20), belong (bv, c).

In addition to Theorems 1.1–1.6 we also need the following results.

Theorem 1.7 Let $A = (a_{nk})$ be a matrix method. Then, $A \in (c_0, cs)$ if and only if

$$\text{all series } \mathfrak{A}_k := \sum_n a_{nk} \text{ are convergent,} \tag{1.36}$$

$$\sum_k \left| \sum_{n=0}^{l} a_{nk} \right| = O(1). \tag{1.37}$$

Theorem 1.8 Let $A = (a_{nk})$ be a matrix method. Then, $A \in (c_0, cs_0)$ if and only if $\mathfrak{A}_k \equiv 0$ and condition (1.37) holds.

Theorem 1.9 Let $A = (a_{nk})$ be a matrix method. Then, $A \in (m, bv) = (c, bv) = (c_0, bv)$ if and only if

$$\left| \sum_{n \in L} \sum_{k \in K} (a_{nk} - a_{n-1,k}) \right| = O(1),$$

where K and L are arbitrary finite subsets of $\mathbf{N}$.

As the proofs of Theorems 1.7–1.9 are rather complicated, we advise the interested reader to consult proofs of these results from [7] and [6]. We also note that the proofs of Theorems 1.1–1.6 can be found in monographs [1–3]. We now present some examples.

Example 1.8 For the method $A_{-1,1}$, $\mathfrak{A}_k \equiv 0$ and

$$\sum_k \left| \sum_{n=0}^{l} a_{nk} \right| = 1;$$

that is, condition (1.37) holds. Thus, by Theorem 1.8, $A_{-1,1} \in (c_0, cs_0)$.

Example 1.9 Let a lower triangular method $A = (a_{nk})$ be defined by

$$a_{nk} := \frac{1}{(n+1)^2(k+1)^2}.$$

Then, clearly the series $\mathfrak{A}_k$ converges to some non-zero number for every k, and

$$\sum_{k=0}^{l} \left| \sum_{n=0}^{l} a_{nk} \right| = \sum_{k=0}^{l} \frac{1}{(k+1)^2} \sum_{n=k}^{l} \frac{1}{(n+1)^2} = O(1);$$

that is, condition (1.37) holds. Thus, by Theorem 1.7, $A \in (c_0, cs)$, but A does not belong to (c_0, cs_0).

Example 1.10 The method A, defined by (1.20), belongs $(m, bv) = (c, bv) = (c_0, bv)$. Indeed,

$$\left| \sum_{n\in L} \sum_{k\in K} (a_{nk} - a_{n-1,k}) \right| = \sum_{n\in L} \frac{1}{n(n+1)} \sum_{k\in K} \frac{1}{k+1}$$

$$\leq \sum_{n\in L} \frac{1}{n(n+1)} \sum_{k=0}^{n} \frac{1}{k+1} = O(1) \sum_{n\in L} \frac{\ln(n+1)}{n(n+1)} = O(1);$$

that is, condition of Theorem 1.9 is fulfilled.

1.3 Excercise

Exercise 1.1 Prove that $A = (a_{nk})$ is conservative, that is, $A \in (c, c)$ if and only if (1.1) holds and the finite limits δ_k and δ exist.
In such a case, prove that

$$\lim_{n\to\infty} (Ax)_n = s\delta + \sum_k (x_k - s)\delta_k, \tag{1.38}$$

$$\lim_{k\to\infty} x_k = s.$$

Hint. Use Theorem 1.1.

Exercise 1.2 Try to prove the Steinhaus theorem without using Theorem 1.2, that is, given a regular matrix, construct a bounded, divergent sequence $x = \{x_k\}$ such that $\{(Ax)_n\}$ diverges.

Exercise 1.3 Prove that $A = (a_{nk}) \in (c_0, c)$ if and only if conditions (1.1) and (1.2) hold.

Hint. Use Theorem 1.1.

Exercise 1.4 Let $A = (a_{nk})$ be a matrix method. Prove that, $A \in (c_0, c_0)$ if and only if conditions (1.1) and (1.2) with $\delta_k \equiv 0$ hold.

Exercise 1.5   Prove that $A = (a_{nk})$ is Sr–Sq regular if and only if condition (1.22) holds and $\delta_k \equiv 1$.

Hint. Use the proof of Theorem 1.5.

Exercise 1.6   Prove that method $A = (a_{nk}) \in (m, m) = (c, m) = (c_0, m)$ if and only if condition (1.1) is satisfied.

Hint. For the proof of the necessity, see the proof of Theorem 1.1.

Exercise 1.7   Prove that $A = (a_{nk}) \in (c, c_0)$ if and only if (1.2), (1.3) are satisfied and

$$\lim_n \sum_k a_{nk} = 0.$$

Exercise 1.8   Prove that a method $A = (a_{nk}) \in (m, c)$ if and only if conditions (1.1), (1.2) are satisfied and

$$\lim_n \sum_k |a_{nk} - \delta_k| = 0.$$

Prove that in this case

$$\lim_n A_n x = \sum_k \delta_k x_k$$

for every $x := (x_k) \in m$.

Hint. We note that this result is a modification of Theorem 1.2.

Exercise 1.9   Prove that $A = (a_{nk}) \in (m, c_0)$ if and only if

$$\lim_n \sum_k |a_{nk}| = 0.$$

Hint. Use Theorem 1.2.

Exercise 1.10   Prove that $A = (a_{nk}) \in (l, m)$ if and only if condition (1.21) holds.

Hint. A matrix A can be considered as a continuous linear operator from l to m. To find the norm of A, use the principle of uniform boundedness.

Exercise 1.11   ([Knopp–Lorentz theorem]) Prove that $A = (a_{nk}) \in (l, l)$ if and only if

$$\sum_n |a_{nk}| = O(1).$$

Hint. The proof is similar to the proof of Theorem 1.4. See also hint of Exercise 1.10.

Exercise 1.12   Prove that $A = (a_{nk}) \in (l, bv)$ if and only if

$$\sum_n |a_{nk} - a_{n-1,k}| = O(1).$$

Hint. Let

$$Y_n := \sum_{k=0}^{n} y_k$$

If $(y_k) \in l$, then $(Y_n) \in bv$, and vice versa, if $(Y_n) \in bv$, then $(y_k) \in l$. Denoting $Y_n := A_n x$ for every $x = (x_k) \in l$, we can say that $(Y_n) \in bv$ for every $x \in l$ if and only if $(y_k) \in l$ for every $x \in l$, where

$$y_n = Y_n - Y_{n-1} = \sum_k (a_{nk} - a_{n-1,k}) x_k.$$

To find conditions for the existence of Ax, use Theorem 1.4. Then, use Exercise 1.10.

Exercise 1.13   Prove that $A = (a_{nk}) \in (bv, bv)$ if and only if

$$\text{series } \sum_k a_{nk} \text{ are convergent,} \tag{1.39}$$

$$\sum_n \left| \sum_{k=0}^{l} (a_{nk} - a_{n-1,k}) \right| = O(1).$$

Moreover, for (bv_0, bv) condition (1.39) is redundant.

Hint. As $e \in bv$, then for finding conditions for the existence of Ax, use Theorem 1.4. Further use Exercise 1.10.

Exercise 1.14   Let A be defined by (1.20). Does $A \in (c, c)$, $A \in (cs, c)$, $A \in (m, c)$, $A \in (l, l)$, $A \in (l, m)$, $A \in (l, bv)$, and $A \in (c_0, cs)$? Is A Sq–Sq or Sr–Sq regular? Why?

Exercise 1.15   Does $A_{-1,1} \in (m, c)$, $A_{-1,1} \in (l, l)$, $A_{-1,1} \in (l, m)$, $A_{-1,1} \in (l, bv)$, and $A_{-1,1} \in (m, bv)$? Why?

Exercise 1.16   Does $Z_{1/2} \in (m, c)$, $Z_{1/2} \in (c_0, c_0)$, $Z_{1/2} \in (l, l)$, $Z_{1/2} \in (l, m)$, $Z_{1/2} \in (l, bv)$, $Z_{1/2} \in (c_0, cs)$, and $Z_{1/2} \in (m, bv)$? Why?

Exercise 1.17   Does $Z^1 \in (m, c)$, $ZZ^1 \in (c_0, c_0)$, $Z^1 \in (l, l)$, $Z^1 \in (l, m)$, $Z^1 \in (l, bv)$, and $Z^1 \in (m, bv)$? Why?

Exercise 1.18 Prove that the method $A = (a_{nk})$, defined by the lower triangular matrix with

$$a_{nk} := \frac{2k}{(n+1)^2},$$

is Sq–Sq regular.

Exercise 1.19 Prove that m is a Banach space with respect to the norm defined by (1.1).

Exercise 1.20 Prove that c, c_0 are closed subspaces of m under the norm defined by (1.1).

Exercise 1.21 If $A = (a_{nk})$, $B = (b_{nk}) \in (c, c)$, prove that $A + B$, $AB \in (c, c)$, where AB denotes the usual matrix product.

Exercise 1.22 Is A regular, if $(Ax)_n = 2x_n - x_{n+1}$ for all n? Why?

Exercise 1.23 Prove that $A = (a_{nk})$ is a Schur matrix if and only if A sums all sequences of 0's and 1's.

Exercise 1.24 ([Mazur–Orlicz theorem]) If a conservative matrix sums a bounded, divergent sequence, prove that it sums an unbounded one too.

References

1 Baron, S.: Vvedenie v teoriyu summiruemosti ryadov (Introduction to the Theory of Summability of Series). Valgus, Tallinn (1977).
2 Boos, J.: Classical and Modern Methods in Summability. Oxford University Press, Oxford (2000).
3 Hardy, G.H.: Divergent Series. Oxford University Press, Oxford (1949).
4 Maddox, I.J.: Elements of Functional Analysis. Cambridge University Press, Cambridge (1970).
5 Powell, R.E. and Shah, S.M.: Summability Theory and Applications. Prentice-Hall of India, Delhi (1988).
6 Snyder, A.K. and Wilansky, A.: Inclusion theorems and semiconservative FK-spaces. Rocky Mt. J. Math. **2**, 595–603 (1972).
7 Zeller, K.: Allgemeine Eigenschaften von Limitierungsverfahren. Math. Z. **53**, 463–487 (1951).

2

Special Summability Methods I

For special methods of summability, standard references are [1, 2].

2.1 The Nörlund Method

In this section, we introduce the Nörlund method.

Definition 2.1 Let $\{p_n\}$ be a sequence of numbers such that $p_n \geq 0$, $n = 1, 2, \ldots$ and $p_0 > 0$. Let

$$P_n = \sum_{k=0}^{n} p_k.$$

Then, the Nörlund method (N, p_n) is defined by the infinite matrix (a_{nk}), where

$$a_{nk} = \begin{cases} \frac{p_{n-k}}{P_n}, & k \leq n; \\ 0, & k > n. \end{cases}$$

Theorem 2.1 The (N, p_n) method is regular if and only if

$$\delta_0 = \lim_{n \to \infty} \frac{p_n}{P_n} = 0. \tag{2.1}$$

Proof: If (N, p_n) is regular, then

$$\lim_{n \to \infty} a_{n0} = 0;$$

that is,

$$\lim_{n \to \infty} \frac{p_n}{P_n} = 0,$$

An Introductory Course in Summability Theory, First Edition. Ants Aasma, Hemen Dutta, and P.N. Natarajan.
© 2017 John Wiley & Sons, Inc. Published 2017 by John Wiley & Sons, Inc.

so that (2.1) holds. Conversely, let (2.1) hold. Then, condition (1.2) with $\delta_k \equiv 0$ is satisfied. Indeed,

$$\frac{p_{n-k}}{P_n} = \frac{p_{n-k}}{P_{n-k}} \frac{P_{n-k}}{P_{n-k+1}} \cdots \frac{P_{n-2}}{P_{n-1}} \frac{P_{n-1}}{P_n},$$

and

$$\lim_{n \to \infty} \frac{p_{n-k}}{P_{n-k}} = \delta_0 = 0,$$

$$\lim_{n \to \infty} \frac{P_{n-k}}{P_{n-k+1}} = \lim_{n \to \infty} \left(1 - \frac{p_{n-k+1}}{P_{n-k+1}} \right) = 1 - \delta_0$$

for every $k \leq n$ by (2.1). Hence,

$$\delta_k = \lim_{n \to \infty} \frac{p_{n-k}}{P_n} = \delta_0 (1 - \delta_0)^k = 0;$$

that is, condition (1.2) with $\delta_k \equiv 0$ is satisfied. As $p_n \geq 0$ for $n = 1, 2, \ldots, p_0 > 0$ and $P_n \geq p_0 > 0$, then

$$\sum_k |a_{nk}| = \sum_{k=0}^{n} a_{nk} = \sum_{k=0}^{n} \frac{p_{n-k}}{P_n} = \frac{1}{P_n} \left(\sum_{k=0}^{n} p_k \right) = \frac{1}{P_n} P_n = 1.$$

Hence,

$$\sup_{n \geq 0} \sum_k |a_{nk}| < \infty.$$

Also,

$$\sum_k a_{nk} = 1,$$

as shown above.

$$\lim_{n \to \infty} \sum_k a_{nk} = 1. \text{ Thus,}$$

Using Theorem 1.1, it now follows that the method (N, p_n) is regular.

Example 2.1 Let $p_0 = 1, p_n = n, n = 1, 2, \ldots$. Now

$$\frac{p_n}{P_n} = \frac{n}{1 + 1 + 2 + \cdots + n} = \frac{n}{1 + \frac{n(n+1)}{2}} = \frac{1}{\frac{1}{n} + \frac{n+1}{2}} \to 0, n \to \infty,$$

so that the (N, p_n) method is regular.

Example 2.2 Let $p_n = 2^n$ for all n. Then,

$$\frac{p_n}{P_n} = \frac{2^n}{1 + 2 + \cdots + 2^n} = \frac{2^n}{2^{n+1} - 1} \to \frac{1}{2}, n \to \infty.$$

By Theorem 2.1, the (N, p_n) method is not regular.

Definition 2.2 Two methods A and B are said to be "consistent" if whenever a sequence $\{x_n\}$ is A-summable to s and B-summable to t, then $s = t$, that is, if the two methods cannot sum the same sequence (or series) to different sums.

Theorem 2.2 Any two regular Nörlund methods (N, p_n) and (N, q_n) are consistent.

Proof: Let

$$r_n = p_0 q_n + p_1 q_{n-1} + \cdots + p_n q_0.$$

Note that $r_n \geq 0$, $n = 1, 2, \ldots, r_0 > 0$. Let $\{\alpha_n\}$, $\{\beta_n\}$, $\{\gamma_n\}$ be the (N, p_n), (N, q_n), (N, r_n)-transforms of a sequence $\{x_n\}$; that is,

$$\alpha_n = \frac{1}{P_n} \sum_{k=0}^{n} p_{n-k} x_k,$$

$$\beta_n = \frac{1}{Q_n} \sum_{k=0}^{n} q_{n-k} x_k,$$

and

$$\gamma_n = \frac{1}{R_n} \sum_{k=0}^{n} r_{n-k} x_k,$$

where, as usual,

$$P_n = \sum_{k=0}^{n} p_k, Q_n = \sum_{k=0}^{n} q_k, R_n = \sum_{k=0}^{n} r_k.$$

Now

$$\gamma_n = \frac{1}{R_n} [r_0 x_n + r_1 x_{n-1} + \cdots + r_n x_0]$$

$$= \frac{1}{R_n} [(p_0 q_0) x_n + (p_0 q_1 + p_1 q_0) x_{n-1}$$

$$+ \cdots + (p_0 q_n + p_1 q_{n-1} + \cdots + p_n q_0) x_0]$$

$$= \frac{1}{R_n} [p_0 (q_0 x_n + q_1 x_{n-1} + \cdots + q_n x_0)$$

$$+ p_1 (q_0 x_{n-1} + q_1 x_{n-2} + \cdots + q_{n-1} x_0) + \cdots + p_n (q_0 x_0)]$$

$$= \frac{1}{R_n} [p_0 Q_n \beta_n + p_1 Q_{n-1} \beta_{n-1} + \cdots + p_n Q_0 \beta_0]$$

$$= \sum_{k} a_{nk} \beta_k,$$

where the matrix (a_{nk}) is defined by

$$
a_{nk} = \begin{cases} \frac{p_{n-k}Q_k}{R_n}, & k \le n; \\ 0, & k > n. \end{cases}
$$

Note that $a_{nk} \ge 0$, for all n, k.

As

$$
R_n = \sum_{k=0}^{n} p_{n-k}Q_k \ne 0,
$$

then

$$
\sum_{k} |a_{nk}| = \sum_{k} a_{nk} = \sum_{k=0}^{n} a_{nk} = \frac{1}{R_n} \sum_{k=0}^{n} p_{n-k}Q_k = \frac{1}{R_n}R_n = 1.
$$

It now follows that

$$
\sup_{n \ge 0} \sum_{k} |a_{nk}| < \infty
$$

and

$$
\lim_{n \to \infty} \sum_{k} a_{nk} = 1.
$$

Also,

$$
R_n = \sum_{k=0}^{n} p_{n-k}Q_k \ge q_0(p_0 + p_1 + \cdots + p_n) = q_0 P_n \ge q_0 P_{n-k},
$$

because $Q_k \ge q_0$. Hence,

$$
a_{nk} = \frac{p_{n-k}Q_k}{R_n} \le \frac{p_{n-k}Q_k}{q_0 P_{n-k}} = \frac{p_{n-k}}{P_{n-k}} \frac{Q_k}{q_0} \to 0, n \to \infty,
$$

using (2.2).

By Theorem 1.1, the infinite matrix (a_{nk}) is regular. So, $\lim_{k \to \infty} \beta_k = t$ implies that $\lim_{n \to \infty} \gamma_n = t$. Similarly, $\lim_{k \to \infty} \alpha_k = s$ implies that $\lim_{n \to \infty} \gamma_n = s$. Thus, $s = t$, completing the proof of the theorem.

Definition 2.3   Given two summability methods A and B, we say that A is included in B (or B includes A), written as

$$
A \subseteq B \quad (\text{or } B \supseteq A),
$$

if, whenever, a sequence $x = \{x_k\}$ is A-summable to s, it is also B-summable to s. A and B are said to be "equivalent" if $A \subseteq B$ and vice versa.

We show that for regular methods (N, p_n) and (N, q_n), the series

$$p(x) = \sum_n p_n x^n, \quad q(x) = \sum_n q_n x^n;$$
$$P(x) = \sum_n P_n x^n, \quad Q(x) = \sum_n Q_n x^n$$

converge if $|x| < 1$. Indeed, by Theorem 2.1, we obtain

$$\lim_n \frac{P_{n-1}}{P_n} = \lim_n \left(1 - \frac{p_n}{P_n} \right) = 1$$

and

$$\lim_n \frac{Q_{n-1}}{Q_n} = 1.$$

Hence, the radius of convergence of power series $P(x)$ and $Q(x)$ are equal 1; that is, the series $P(x)$ and $Q(x)$ are convergent for $|x| < 1$.
As

$$p(x) = \sum_n (P_n - P_{n-1})x^n = \sum_n P_n x^n - x \sum_n P_{n-1} x^{n-1} = (1 - x)P(x)$$

and

$$q(x) = (1 - x)Q(x),$$

then for $|x| < 1$ the series $p(x)$, $q(x)$ also are convergent and

$$\frac{q(x)}{p(x)} = \frac{Q(x)}{P(x)}, \quad \frac{p(x)}{q(x)} = \frac{P(x)}{Q(x)}.$$

Now it is easy to see that the series

$$k(x) = \sum_n k_n x^n = \frac{q(x)}{p(x)} = \frac{Q(x)}{P(x)},$$

$$h(x) = \sum_n h_n x^n = \frac{p(x)}{q(x)} = \frac{P(x)}{Q(x)},$$

converge for $|x| < 1$, and

$$k_0 p_n + k_1 p_{n-1} + \cdots + k_n p_0 = q_n, \quad k_0 P_n + k_1 P_{n-1} + \cdots + k_n P_0 = Q_n;$$
$$h_0 q_n + h_1 q_{n-1} + \cdots + h_n q_0 = p_n, \quad h_0 Q_n + h_1 Q_{n-1} + \cdots + h_n Q_0 = P_n.$$

We now prove an inclusion theorem for Nörlund methods.

Theorem 2.3 (see [1, Theorem 19]) Let (N, p_n), (N, q_n) be regular Nörlund methods. Then,

$$(N, p_n) \subseteq (N, q_n)$$

if and only if

$$|k_0|P_n + |k_1|P_{n-1} + \cdots + |k_n|P_0 \le HQ_n, H > 0, \text{ for all } n, \qquad (2.2)$$

which we also write as

$$\sum_{j=0}^{n} |k_{n-j}|P_j = O(Q_n), \quad n \to \infty,$$

and

$$\frac{k_n}{Q_n} \to 0, \quad n \to \infty, \qquad (2.3)$$

which is also written as

$$k_n = o(Q_n), \quad n \to \infty.$$

Proof: Let $\{\alpha_n\}$, $\{\beta_n\}$ be the (N, p_n), (N, q_n)-transforms of the sequence $s = \{s_n\}$ respectively. Then,

$$\alpha_n = \frac{1}{P_n} \sum_{k=0}^{n} p_{n-k} s_k,$$

$$\beta_n = \frac{1}{Q_n} \sum_{k=0}^{n} q_{n-k} s_k.$$

Let $s(x) = \Sigma_n s_n x^n$. Now, we have

$$\sum_n Q_n \beta_n x^n = \sum_n (q_0 s_n + q_1 s_{n-1} + \cdots + q_n s_0) x^n$$

$$= q(x) s(x),$$

for small x. Similarly,

$$\sum_n P_n \alpha_n x^n = p(x) s(x),$$

for small x so that

$$\frac{\sum\limits_n Q_n \beta_n x^n}{\sum\limits_n P_n \alpha_n x^n} = \frac{q(x) s(x)}{p(x) s(x)} = \frac{q(x)}{p(x)} = k(x).$$

Thus,

$$\sum_n Q_n \beta_n x^n = k(x) \left(\sum_n P_n \alpha_n x^n \right)$$

$$= \left(\sum_n k_n x^n \right) \left(\sum_n P_n \alpha_n x^n \right),$$

from which it follows that

$$Q_n \beta_n = k_0 P_n \alpha_n + k_1 P_{n-1} \alpha_{n-1} + \cdots + k_n P_0 \alpha_0,$$

so

$$\beta_n = \sum_r a_{nr} \alpha_r,$$

where the infinite matrix (a_{nr}) is defined by

$$a_{nr} = \begin{cases} \frac{k_{n-r} P_r}{Q_n}, & r \leq n; \\ 0, & r > n. \end{cases}$$

Hence, we obtain

$$\sum_r a_{nr} = \sum_{r=0}^{n} a_{nr} = \sum_{r=0}^{n} \frac{k_{n-r} P_r}{Q_n} = \frac{1}{Q_n} \sum_{r=0}^{n} k_{n-r} P_r = \frac{1}{Q_n} Q_n = 1,$$

since

$$Q_n = \sum_{r=0}^{n} k_{n-r} P_r \neq 0.$$

This implies that

$$\lim_{n \to \infty} \sum_r a_{nr} = 1.$$

So (a_{nr}) is regular by Theorem 1.1 if and only if (1.2) and (1.3) hold. Condition (1.2) is equivalent to the condition

$$\sup_{n \geq 0} \sum_{r=0}^{n} |k_{n-r}| \frac{P_r}{Q_n} < \infty,$$

or

$$\sum_{r=0}^{n} |k_{n-r}| P_r \leq H Q_n, H > 0;$$

that is, condition (1.2) is equivalent to (2.2). Moreover, condition (1.3) is equivalent to

$$\lim_{n \to \infty} \frac{k_{n-r} P_r}{Q_n} = 0, \quad \text{for all } r,$$

$$\text{i.e.,} \quad \lim_{n \to \infty} \frac{k_{n-r}}{Q_n} = 0, \quad \text{for all } r,$$

$$\text{i.e.,} \quad \lim_{n \to \infty} \frac{k_{n-r}}{Q_{n-r}} \cdot \frac{Q_{n-r}}{Q_n} = 0, \quad \text{for all } r,$$

$$\text{i.e.,} \quad \lim_{n \to \infty} \frac{k_{n-r}}{Q_{n-r}} = 0, \quad \text{for all } r,$$

since

$$\lim_{n\to\infty} \frac{Q_{n-r}}{Q_n} = 1.$$

Thus, (1.3) is equivalent to (2.3). Consequently,

$$(N, p_n) \subseteq (N, q_n),$$

if and only if (2.2) and (2.3) hold.

We now prove an equivalence theorem.

Theorem 2.4   (see [1, Theorem 21]) Regular Nörlund methods (N, p_n), (N, q_n) are equivalent if and only if

$$\sum_n |k_n| < \infty \text{ and } \sum_n |h_n| < \infty. \tag{2.4}$$

Proof: Necessity. $p_0, q_0 > 0$ imply $k_0, h_0 > 0$. As

$$(N, p_n) \subseteq (N, q_n),$$

it follows from (2.2) that

$$k_0 P_n \leq H Q_n,$$

i.e., $\dfrac{P_n}{Q_n} \leq \dfrac{H}{k_0}$, $n = 0, 1, 2, \cdots$.

Thus, $\{P_n/Q_n\}$ is bounded. Similarly, $\{Q_n/P_n\}$ is also bounded.
Using (2.2), we have

$$|k_0| + |k_1| \frac{P_{n-1}}{P_n} + \cdots + |k_r| \frac{P_{n-r}}{P_n} \leq H \frac{Q_n}{P_n}, \ r \leq n.$$

Fixing r and taking the limit as $n \to \infty$, we see that

$$|k_0| + |k_1| + \cdots + |k_r| \leq H \overline{\lim_{n\to\infty}} \left(\frac{Q_n}{P_n} \right),$$

since

$$\lim_{n\to\infty} \frac{P_{n-j}}{P_n} = 1, \ j = 0, 1, 2, \ldots, r.$$

It now follows that $\Sigma_n |k_n| < \infty$. Similarly, we have $\Sigma_n |h_n| < \infty$, completing the proof of the necessity part.

Sufficiency. Let (2.4) hold. Since $\Sigma_n |k_n| < \infty$,

$$k_n \to 0, \ n \to \infty.$$

Also,

$$Q_n \geq q_0, \text{ for all } n.$$

So,

$$\frac{1}{Q_n} \leq \frac{1}{q_0}, \text{ for all } n.$$

Thus,

$$\left|\frac{k_n}{Q_n}\right| = \frac{|k_n|}{Q_n} \leq \frac{|k_n|}{q_0} \to 0, \ n \to \infty,$$

so that

$$\frac{k_n}{Q_n} \to 0, \ n \to \infty.$$

We also have

$$P_n = Q_0 h_n + Q_1 h_{n-1} + \cdots + Q_n h_0 \leq Q_n \sum_n |h_n|,$$

from which it follows that

$$P_n|k_0| + P_{n-1}|k_1| + \cdots + P_0|k_n| \leq \left(\sum_n |k_n|\right)\left(\sum_n |h_n|\right) Q_n.$$

So (2.2) holds with

$$H = \left(\sum_n |k_n|\right)\left(\sum_n |h_n|\right).$$

We have already proved that $k_n/Q_n \to 0$, $n \to \infty$, so that (2.3) holds. In view of Theorem 2.3,

$$(N, p_n) \subseteq (N, q_n).$$

Similarly, we can prove that

$$(N, q_n) \subseteq (N, p_n),$$

so that (N, p_n) and (N, q_n) are equivalent.

More inclusion theorems on Nörlund methods appear in [1], pp. 69–70.

2.2 The Weighted Mean Method

Definition 2.4 Let $\{p_n\}$ be a sequence of nonnegative numbers such that $p_0 > 0$, that is, $p_n \geq 0$, $n = 1, 2, \ldots$ and $p_0 > 0$. The Weighted Mean method $(\overline{N}, p_n)$ is defined by the infinite matrix (a_{nk}), where (a_{nk}) is defined by

$$a_{nk} = \begin{cases} \frac{p_k}{P_n}, & k \leq n; \\ 0, & k > n. \end{cases}$$

Theorem 2.5 The $(\overline{N}, p_n)$ method is regular if and only if

$$\lim_{n \to \infty} P_n = \infty. \tag{2.5}$$

Proof: As $P_n \neq 0$, then

$$\sum_k |a_{nk}| = \sum_k a_{nk} = \sum_{k=0}^{n} a_{nk} = \sum_{k=0}^{n} \frac{p_k}{P_n} = \frac{1}{P_n} \sum_{k=0}^{n} p_k = \frac{1}{P_n} P_n = 1.$$

It now follows that $\sup_{n \geq 0} \Sigma_k |a_{nk}| < \infty$ and

$$\lim_{n \to \infty} \sum_k a_{nk} = 1.$$

It is easy to see that condition $\lim_{n \to \infty} a_{nk} = 0$ is satisfied if and only if condition (2.5) holds. Therefore, by Theorem 1.1, $(\overline{N}, p_n)$ is regular if and only if condition (2.5) holds.

The following is a limitation theorem on sequences that are $(\overline{N}, p_n)$-summable.

Theorem 2.6 (Limitation theorem) If $p_n > 0$, for all n and $\{s_n\}$ is $(\overline{N}, p_n)$ summable to s, then

$$s_n - s = o\left(\frac{P_n}{p_n}\right), \quad n \to \infty. \tag{2.6}$$

Proof: Let $\{t_n\}$ be the $(\overline{N}, p_n)$-transform of $\{s_n\}$, so that

$$t_n = \frac{\sum_{k=0}^{n} p_k s_k}{P_n}.$$

By hypothesis, $\lim_{n \to \infty} t_n = s$. Note that

$$p_n s_n = P_n t_n - P_{n-1} t_{n-1},$$

for all n, $P_{-1} = t_{-1} = 0$.
Now, as $n \to \infty$, we obtain

$$\frac{p_n}{P_n}(s_n - s) = \frac{1}{P_n}(P_n t_n - P_{n-1} t_{n-1}) - \frac{p_n}{P_n} s$$

$$= (t_n - s) - \frac{P_{n-1}}{P_n}(t_{n-1} - s) + s - \frac{P_{n-1}}{P_n} s - \frac{p_n}{P_n} s$$

$$= (t_n - s) - \frac{P_{n-1}}{P_n}(t_{n-1} - s) + s\left(1 - \frac{P_n}{P_n}\right) \to 0,$$

because

$$\lim_{n \to \infty} t_n = s \text{ and } P_{n-1}/P_n \leq 1.$$

This proves that

$$s_n - s = o\left(\frac{P_n}{p_n}\right), n \to \infty.$$

Note that Theorem 2.6 puts some restriction on the terms of a sequence $\{s_n\}$ that is $(\overline{N}, p_n)$-summable.

We now prove an inclusion theorem for Weighted Mean methods.

Theorem 2.7 Let $p_n, q_n > 0$, $n = 0, 1, 2, \dots$ and $(\overline{N}, p_n)$, $(\overline{N}, q_n)$ be regular. If

$$\frac{q_{n+1}}{q_n} \leq \frac{p_{n+1}}{p_n} \quad \left(\text{or} \frac{p_{n+1}}{p_n} \leq \frac{q_{n+1}}{q_n}\right) \tag{2.7}$$

and

$$\frac{P_n q_n}{p_n Q_n} = O(1), n \to \infty, \tag{2.8}$$

then

$$(\overline{N}, p_n) \subseteq (\overline{N}, q_n).$$

Proof: Let $\{t_n\}$, $\{u_n\}$ be the $(\overline{N}, p_n)$, $(\overline{N}, q_n)$-transforms of the sequence $\{s_n\}$;

$$t_n = \frac{1}{P_n} \sum_{k=0}^{n} p_k s_k,$$

$$u_n = \frac{1}{Q_n} \sum_{k=0}^{n} q_k s_k.$$

Then,

$$p_0 s_0 = P_0 t_0,$$
$$p_n s_n = P_n t_n - P_{n-1} t_{n-1}, n = 1, 2, \dots.$$

So,

$$u_n = \frac{1}{Q_n}[q_0 s_0 + q_1 s_1 + \cdots + q_n s_n]$$

$$= \frac{1}{Q_n}\left[q_0 \frac{P_0 t_0}{P_0} + q_1 \left(\frac{P_1 t_1 - P_0 t_0}{P_1}\right) \right.$$

$$\left. + \cdots + q_n \left(\frac{P_n t_n - P_{n-1} t_{n-1}}{P_n}\right)\right]$$

$$= \sum_k a_{nk} t_k,$$

where the infinite matrix (a_{nk}) is defined by

$$a_{nk} = \begin{cases} \left(\frac{q_k}{P_k} - \frac{q_{k+1}}{P_{k+1}} \right) \frac{P_k}{Q_n}, & k < n; \\ \frac{q_n P_n}{P_n Q_n}, & k = n; \\ 0, & k > n. \end{cases}$$

Now, if $s_n = 1$, for all n,

$$t_n = 1, u_n = 1,$$

so that

$$\sum_k a_{nk} = 1,$$

for all n, and consequently,

$$\lim_{n\to\infty} \sum_k a_{nk} = 1.$$

Since $(\overline{N}, q_n)$ is regular,

$$\lim_{n\to\infty} Q_n = \infty,$$

so that, for fixed k,

$$\lim_{n\to\infty} a_{nk} = 0.$$

It remains to prove that

$$\sup_{n\geq 0} \sum_k |a_{nk}| < \infty.$$

If

$$\frac{q_{n+1}}{q_n} \leq \frac{p_{n+1}}{p_n},$$

then $a_{nk} \geq 0$. Hence,

$$\sum_k |a_{nk}| = \sum_k a_{nk} = 1,$$

from which it follows that

$$\sup_{n\geq 0} \sum_k |a_{nk}| < \infty.$$

If

$$\frac{p_{n+1}}{p_n} \leq \frac{q_{n+1}}{q_n},$$

then $a_{nk} \le 0$, except when $k = n$, in which case $a_{n,n} > 0$. Hence,

$$\sum_k |a_{nk}| = -\sum_{k=0}^{n-1} a_{nk} + \frac{q_n P_n}{p_n Q_n},$$

or

$$1 = \sum_k a_{nk} = \sum_{k=0}^{n-1} a_{nk} + \frac{q_n P_n}{p_n Q_n}.$$

Consequently,

$$\sum_k |a_{nk}| = \left(\frac{q_n P_n}{p_n Q_n} - 1 \right) + \frac{q_n P_n}{p_n Q_n} = 2\frac{q_n P_n}{p_n Q_n} - 1 = O(1), n \to \infty$$

by (2.8). So

$$\sup_{n \ge 0} \sum_k |a_{nk}| < \infty$$

in this case too. Thus, (a_{nk}) is regular and so $\lim_{k \to \infty} t_k = s$ implies that $\lim_{n \to \infty} u_n = s$. In other words,

$$(\overline{N}, p_n) \subseteq (\overline{N}, q_n),$$

completing the proof of the theorem.

When $\Sigma_n p_n$ diverges rapidly, the method $(\overline{N}, p_n)$ becomes trivial, in the sense that $(\overline{N}, p_n)$ sums only convergent sequences. The following theorem proves our claim.

Theorem 2.8 If $P_{n+1}/P_n \ge 1 + \delta > 1$, then $\{s_n\}$ cannot be $(\overline{N}, p_n)$ summable unless it is convergent.

Proof: Let $\{t_n\}$ be the $(\overline{N}, p_n)$-transform of the sequence $\{s_n\}$; that is,

$$t_n = \frac{1}{P_n} \left(\sum_{k=0}^{n} p_k s_k \right),$$

so that

$$s_n = \frac{P_n t_n - P_{n-1} t_{n-1}}{p_n} = \sum_k a_{nk} t_k,$$

with $P_{-1} = t_{-1} = 0$, where (a_{nk}) is defined by

$$a_{nk} = \begin{cases} -\frac{P_{n-1}}{p_n}, & k = n - 1; \\ \frac{P_n}{p_n}, & k = n; \\ 0, & \text{otherwise.} \end{cases}$$

It is clear that

$$\lim_{n \to \infty} a_{nk} = 0, \text{ for all } k.$$

Further,

$$\sum_k a_{nk} = -\frac{P_{n-1}}{p_n} + \frac{P_n}{p_n} = \frac{P_n - P_{n-1}}{p_n} = \frac{p_n}{p_n} = 1,$$

since $p_n \neq 0$. Hence,

$$\lim_{n \to \infty} \sum_k a_{nk} = 1.$$

Also we have

$$\sum_k |a_{nk}| = \frac{P_n + P_{n-1}}{p_n} = \frac{p_n + 2P_{n-1}}{p_n} = 2\frac{P_{n-1}}{p_n} + 1.$$

By hypothesis,

$$\frac{P_{n+1}}{P_n} = \frac{P_n + p_{n+1}}{P_n} = 1 + \frac{p_{n+1}}{P_n} \geq 1 + \delta;$$

that is,

$$\frac{p_{n+1}}{P_n} \geq \delta$$

or

$$p_{n+1} \geq \delta P_n.$$

Consequently,

$$\sum_k |a_{nk}| \leq 1 + \frac{2}{\delta},$$

from which we can conclude that

$$\sup_{n \geq 0} \sum_k |a_{nk}| \leq 1 + \frac{2}{\delta} < \infty.$$

Thus, the infinite matrix (a_{nk}) is regular. Consequently, $\lim_{n \to \infty} t_n = s$ implies that $\lim_{n \to \infty} s_n = s$. Proof of the theorem is now complete.

2.3 The Abel Method and the $(C, 1)$ Method

In this section, we first introduce the Abel method. We incidentally note that the Abel method cannot be defined by an infinite matrix so that there are "nonmatrix summability methods." An Abel method can also be regarded as a semicontinuous method. Poisson used this method in the summation of Fourier series and the method is sometimes attributed to Poisson.

Definition 2.5 A sequence $\{a_n\}$ is said to be Abel summable, written as (A) summable to L if

$$\lim_{x\to 1-}(1-x)\sum_k a_k x^k \text{ exists finitely and } = L,$$

where, by writing

$$\lim_{x\to 1-} f(x) = M,$$

we mean the following: given $\epsilon > 0$, there exists $\delta = \delta(\epsilon) > 0$ such that

$$|f(x) - M| < \epsilon, \text{whenever, } 1-\delta < x < 1.$$

The following result establishes the regularity of the Abel method (A).

Theorem 2.9 If $\{a_n\}$ converges to L, then $\{a_n\}$ is Abel summable to L.

Proof: Let

$$f(x) = (1-x)\sum_k a_k x^k \text{for } |x| < 1, \text{i.e., } -1 < x < 1.$$

Let $0 < \epsilon < 1$ be given. We can now choose a positive integer N such that

$$|a_n - L| < \frac{\epsilon}{2}, n > N. \tag{2.9}$$

Let $M = \max_{0\le i\le N}|a_i - L|$. Note that

$$(1-x)\left(\sum_k x^k\right) = (1-x)\frac{1}{1-x}$$

$$= 1, \text{if } |x| < 1.$$

Let

$$\delta = \frac{\epsilon}{2(M+1)(N+1)}.$$

Then, if $x \in (1-\delta, 1)$, by (2.9), we get

$$|f(x) - L| = \left|(1-x)\sum_k a_k x^k - L\right|$$

$$= \left|(1-x)\sum_k a_k x^k - (1-x)\left(\sum_k Lx^k\right)\right|$$

$$= \left| (1-x) \sum_k (a_k - L)x^k \right|$$

$$= \left| (1-x) \left\{ \sum_{k=0}^{N} (a_k - L)x^k + \sum_{k=N+1}^{\infty} (a_k - L)x^k \right\} \right|$$

$$\leq \left| (1-x) \sum_{k=0}^{N} (a_k - L)x^k \right| + \left| (1-x) \sum_{k=N+1}^{\infty} (a_k - L)x^k \right|$$

$$< (1-x)(N+1)M + \frac{\epsilon}{2} < (N+1)M\delta + \frac{\epsilon}{2}$$

$$= (N+1)M \frac{\epsilon}{2(N+1)(M+1)} + \frac{\epsilon}{2} < \frac{\epsilon}{2} + \frac{\epsilon}{2} = \epsilon.$$

Thus, $\lim_{x \to 1-} f(x)$ exists finitely with

$$\lim_{x \to 1-} (1-x) \sum_k a_k x^k = L;$$

that is, $\{a_n\}$ is Abel summable to L. This implies that the Abel method (A) is regular.

We show now that the converse does not hold. We consider the following example.

Example 2.3 Let $a_n = 1 + (-1)^n$, for all n. That is, $\{a_n\} = \{2, 0, 2, 0, \cdots\}$. It is clear that $\{a_n\}$ diverges. However,

$$\lim_{x \to 1-} (1-x) \sum_k a_k x^k = \lim_{x \to 1-} 2 \sum_k (-1)^k x^k = 2 \lim_{x \to 1-} \left(\frac{1}{1+x} \right) = 2 \cdot \frac{1}{2} = 1,$$

which shows that $\{a_n\}$ is (A) summable to 1.

We now proceed to define $(C, 1)$ convergence.

Definition 2.6 A sequence $\{a_n\}$ is said to be $(C, 1)$ summable to L if

$$\lim_{n \to \infty} \frac{1}{n+1} \sum_{k=0}^{n} a_k \text{ exists finitely and equals } L.$$

From the above definition, we see that the notion of $(C, 1)$ summability is to take the arithmetic mean of the terms of the given sequence and study the convergence of these means.

It is clear that the $(C, 1)$ method of summability is defined by the infinite matrix (a_{nk}), where

$$a_{nk} = \begin{cases} \frac{1}{n+1}, & k \leq n; \\ 0, & k > n. \end{cases}$$

The following result is easily proved.

Theorem 2.10   The $(C, 1)$ method is regular.

We now give an example of a nonconvergent sequence that is $(C, 1)$-summable.

Example 2.4   Consider the sequence $\left\{ \frac{1}{2}, -\frac{1}{2}, 0, \frac{1}{2}, -\frac{1}{2}, 0, \frac{1}{2}, -\frac{1}{2}, 0, \cdots \right\}$, which is clearly not convergent. One can easily verify that this sequence is $(C, 1)$ summable to 0.

Theorem 2.11 gives a necessary condition for $(C, 1)$ summability.

Theorem 2.11   (Limitation theorem) If $\Sigma_k a_k$ is $(C, 1)$-summable, then

$$s_n = o(n), a_n = o(n), n \to \infty,$$

where

$$s_n = \sum_{k=0}^{n} a_k, \text{ for all } n.$$

Proof: Let

$$t_n = \frac{1}{n+1} \sum_{k=0}^{n} s_k, \text{ for all } n,$$

so that

$$s_n = (n+1)t_n - nt_{n-1}.$$

By hypothesis, $\{t_n\}$ converges to L (say). Now

$$\frac{s_n}{n} = \left(1 + \frac{1}{n} \right) t_n - t_{n-1}$$

$$= (t_n - t_{n-1}) + \frac{t_n}{n}.$$

Since $\{t_n\}$ converges, it is bounded, that is, $|t_n| \leq M$, $n = 0, 1, 2, \cdots$, $M > 0$. Thus,

$$\left| \frac{t_n}{n} \right| \leq \frac{M}{n} \to 0, n \to \infty,$$

and consequently,

$$\lim_{n\to\infty} \frac{s_n}{n} = L - L + 0 = 0,$$

which implies that

$$s_n = o(n), n \to \infty.$$

Further,

$$a_n = s_n - s_{n-1},$$

and

$$\frac{a_n}{n} = \frac{s_n}{n} - \frac{s_{n-1}}{n} = \frac{s_n}{n} - \frac{s_{n-1}}{n-1}\left(\frac{n-1}{n}\right) = \frac{s_n}{n} - \frac{s_{n-1}}{n-1}\left(1 - \frac{1}{n}\right).$$

So

$$\lim_{n\to\infty} \frac{a_n}{n} = 0 - 0(1) = 0,$$

implying that

$$a_n = o(n), n \to \infty.$$

Example 2.5   Consider the series

$$1 - 2 + 3 - 4 + \cdots + (-1)^{n-1}n + \cdots.$$

For this series,

$$a_n \neq o(n), n \to \infty,$$

and so this series cannot be $(C, 1)$-summable.

The following is an instance where $(C, 1)$-summability is applied.

Theorem 2.12   If $\Sigma_k a_k = A$, $\Sigma_k b_k = B$, then the Cauchy product series $\Sigma_k c_k$ is $(C, 1)$-summable to AB.

Proof: We recall that

$$c_n = \sum_{k=0}^{n} a_k b_{n-k}$$

for all n. Let

$$A_n = \sum_{k=0}^{n} a_k, B_n = \sum_{k=0}^{n} b_k, s_n = \sum_{k=0}^{n} c_k,$$

$$E_n = \frac{1}{n+1} \sum_{k=0}^{n} s_k,$$

for all n. Now

$$S_n = \sum_{k=0}^{n} c_k = \sum_{k=0}^{n} \left(\sum_{j=0}^{k} a_j b_{k-j} \right) = \sum_{j=0}^{n} \left(\sum_{k=j}^{n} a_j b_{k-j} \right)$$

$$= \sum_{j=0}^{n} a_j \left(\sum_{k=j}^{n} b_{k-j} \right) = \sum_{j=0}^{n} a_j B_{n-j}.$$

Also,

$$E_n = \frac{1}{n+1} \sum_{k=0}^{n} s_k = \frac{1}{n+1} \sum_{k=0}^{n} \left(\sum_{j=0}^{k} a_j B_{k-j} \right)$$

$$= \frac{1}{n+1} \sum_{k=0}^{n} \left(\sum_{j=0}^{k} a_{k-j} B_j \right)$$

$$= \frac{1}{n+1} \sum_{j=0}^{n} \left(\sum_{k=j}^{n} a_{k-j} B_j \right) = \frac{1}{n+1} \sum_{j=0}^{n} B_j \left(\sum_{k=j}^{n} a_{k-j} \right)$$

$$= \frac{1}{n+1} \left(\sum_{j=0}^{n} B_j A_{n-j} \right) = \frac{1}{n+1} \left[\sum_{j=0}^{n} B_j \{(A_{n-j} - A) + A\} \right]$$

$$= \frac{1}{n+1} \sum_{j=0}^{n} B_j (A_{n-j} - A) + A \left[\frac{1}{n+1} \sum_{j=0}^{n} B_j \right]. \tag{2.10}$$

Since $\{B_n\}$ converges to B, it is $(C, 1)$-summable to B, using Theorem 2.10, so that

$$\lim_{n \to \infty} \frac{1}{n+1} \sum_{j=0}^{n} B_j = B.$$

Thus,

$$\lim_{n \to \infty} E_n = AB$$

if and only if

$$\lim_{n \to \infty} \frac{1}{n+1} \sum_{j=0}^{n} B_j (A_{n-j} - A) = 0.$$

Take any $\epsilon > 0$. Since $\{B_n\}$, $\{A_n - A\}$ converge, they are bounded and so for some $M > 0$,

$$|B_n| \leq M, |A_n - A| \leq M \tag{2.11}$$

for all n.

Since $\lim_{n\to\infty}(A_n - A) = 0$, there exists a positive integer N_1 such that

$$|A_n - A| < \frac{\epsilon}{2M}, \ n > N_1. \tag{2.12}$$

Since

$$\lim_{x_n\to\infty} \frac{N_1 + 1}{n} = 0, \tag{2.13}$$

we can choose a positive integer N_2 such that

$$\frac{N_1 + 1}{N_2} < \frac{\epsilon}{2M^2}. \tag{2.14}$$

Let $N = \max(N_1, N_2)$. So, if $n > N$, we obtain, using (2.11), (2.12),

$$\left| \frac{1}{n+1} \sum_{j=0}^{n} B_j(A_{n-j} - A) \right|$$

$$= \left| \frac{1}{n+1} \left\{ \sum_{j=0}^{N_1} B_j(A_{n-j} - A) + \sum_{j=N_1+1}^{n} B_j(A_{n-j} - A) \right\} \right|$$

$$\leq \frac{1}{n+1} \sum_{j=0}^{N_1} |B_j||A_{n-j} - A| + \frac{1}{n+1} \sum_{j=N_1+1}^{n} |B_j||A_{n-j} - A|$$

$$< \frac{1}{n+1}(N_1 + 1)M^2 + \frac{1}{n+1}\frac{\epsilon}{2M}(n - N_1)M$$

$$= \frac{(N_1 + 1)M^2}{n+1} + \frac{n - N_1}{n+1}\frac{\epsilon}{2} < \frac{(N_1 + 1)M^2}{n+1} + \frac{\epsilon}{2},$$

because

$$\frac{n - N_1}{n+1} < 1.$$

If $n > N$, then

$$n + 1 > N_2, \ 1/(n+1) < 1/N_2, \ (N_1 + 1)/(n+1) < (N_1 + 1)/N_2 < \epsilon/2M^2,$$

by (2.14). Thus, if $n > N$, then

$$\left| \frac{1}{n+1} \sum_{j=0}^{n} B_j(A_{n-j} - A) \right| < \frac{\epsilon}{2M^2}M^2 + \frac{\epsilon}{2} = \frac{\epsilon}{2} + \frac{\epsilon}{2} = \epsilon.$$

This proves that

$$\lim_{n\to\infty} \frac{1}{n+1} \left\{ \sum_{j=0}^{n} B_j(A_{n-j} - A) \right\} = 0,$$

from which we conclude that

$$\lim_{n \to \infty} E_n = AB;$$

that is, the Cauchy product series $\sum_k c_k$ is $(C, 1)$-summable to AB.

We have an important consequence of Theorem 2.12.

Corollary 2.1 If $\sum_k a_k = A$, $\sum_k b_k = B$, and $\sum_k c_k = C$, then $C = AB$.

Proof: By Theorem 2.12, $\sum_k c_k$ is $(C, 1)$-summable to AB. Since $\sum_k c_k$ converges to C and $(C, 1)$-summability is regular, using Theorem 2.10, $\sum_k c_k$ is $(C, 1)$-summable to C. It now follows that $C = AB$, completing the proof.

We conclude the present section with a result comparing Abel summability and $(C, 1)$ summability.

Theorem 2.13 If $\{a_n\}$ is $(C, 1)$-summable to L, then it is Abel-summable to L but not conversely. Symbolically, $(C, 1) \subsetneq (A)$.

Proof: Let

$$s_n = \frac{1}{n+1} \sum_{k=0}^{n} a_k,$$

and

$$f(x) = (1 - x) \sum_k a_k x^k.$$

Now

$$(n + 1)s_n = \sum_{k=0}^{n} a_k,$$

so that

$$a_k = (k + 1)s_k - ks_{k-1}, s_{-1} = 0.$$

Hence,

$$f(x) = (1 - x) \sum_k a_k x^k$$

$$= (1 - x) \left[a_0 + \sum_{k=1}^{\infty} \{(k + 1)s_k - ks_{k-1}\} x^k \right].$$

Since $\Sigma_k x^{k+1}$ has radius of convergence 1, the series $\Sigma_k(k+1)x^k$ has radius of convergence 1 and so $\Sigma_k(k+1)s_k x^k$ and $\Sigma_k k s_{k-1} x^k$ each has radius of convergence at least 1, since $\{s_k\}$ converges to a finite limit L. Thus, the power series

$$\sum_{k=1}^{\infty} \{(k+1)s_k - ks_{k-1}\}x^k$$

has the radius of convergence at least 1 and so, the power series

$$(1-x)\sum_k a_k x^k$$

converges for $|x| < 1$.

At this point, we claim that

$$\lim_{x\to 1-}(1-x)\sum_k a_k x^k = L.$$

Now

$$\frac{1}{(1-x)^2}f(x) = \frac{1}{(1-x)^2}(1-x)\sum_k a_k x^k$$

$$= \frac{1}{1-x}\sum_k a_k x^k$$

$$= \left(\sum_k x^k\right)\left(\sum_k a_k x^k\right)$$

$$= \sum_k (a_0 + a_1 + \cdots + a_k)x^k$$

$$= \sum_k (k+1)s_k x^k.$$

We know that

$$\frac{1}{(1-x)^2} = \sum_k (k+1)x^k.$$

Hence, for $0 < x < 1$, we get

$$|f(x) - L| = \left|(1-x)^2\sum_k (k+1)s_k x^k - L(1-x)^2\sum_k (k+1)x^k\right|$$

$$= \left|(1-x)^2\sum_k (k+1)(s_k - L)x^k\right|$$

$$\leq (1-x)^2\sum_k (k+1)|s_k - L|x^k.$$

Since $\lim_{n\to\infty}s_n = L$, given an $\epsilon > 0$, there exists a positive integer N, such that

$$|s_n - L| < \frac{\epsilon}{2}, n \geq N.$$

Also $\{s_n\}$ is bounded, and so

$$|s_n| \leq M, M > 0, \text{ for all } n.$$

Let

$$\delta = \min\left[\frac{1}{2}, \left\{\frac{\epsilon}{4(M+1)(N+1)^2}\right\}^{\frac{1}{2}}\right].$$

If $1 - \delta < x < 1$, then

$$|f(x) - L| \leq (1-x)^2 \left[\sum_{k=0}^{N-1}(k+1)|s_k - L|x^k + \sum_{k=N}^{\infty}(k+1)|s_k - L|x^k\right]$$

$$< \delta^2 N(2M)\sum_{k=0}^{N-1}x^k + (1-x)^2\frac{\epsilon}{2}\sum_k(k+1)x^k$$

$$< \delta^2 N^2(2M) + (1-x)^2\frac{\epsilon}{2}\frac{1}{(1-x)^2}$$

$$\leq \frac{\epsilon}{4(M+1)(N+1)^2}2N^2M + \frac{\epsilon}{2} < \frac{\epsilon}{2} + \frac{\epsilon}{2} = \epsilon.$$

Thus,

$$\lim_{x \to 1-} f(x) = L \text{ or } \lim_{x \to 1-}(1-x)\sum_k a_k x^k = L;$$

that is, $\{a_n\}$ is Abel-summable to L, completing the proof.

To prove that the converse of the theorem does not hold, consider the following example.

Example 2.6   Consider the sequence $\{a_n\}$, where

$$a_n = \begin{cases} (k+1), & \text{if } n = 2k; \\ -(k+1), & \text{if } n = 2k+1. \end{cases}$$

Let

$$s_n = \frac{1}{n+1}\sum_{k=0}^{n}a_k.$$

Then,

$$s_n = \begin{cases} \frac{k+1}{2k+1}, & \text{if } n = 2k; \\ 0, & \text{if } n = 2k+1. \end{cases}$$

Since $\{s_{2k}\}$ converges to $1/2$ and $\{s_{2k+1}\}$ converges to 0, $\{s_n\}$ does not converge. Thus, $\{a_n\}$ is not $(C, 1)$-summable.

If $|x| < 1$, then

$$\sum_k a_k x^k = 1 - x + 2x^2 - 2x^3 + 3x^4 - 3x^5 + \cdots$$

$$= (1 - x) + 2x^2(1 - x) + 3x^4(1 - x) + \cdots$$
$$= (1 - x)[1 + 2x^2 + 3x^4 + \cdots]$$
$$= (1 - x)\frac{1}{(1 - x^2)^2}$$

and so,

$$(1 - x)\sum_k a_k x^k = (1 - x)^2 \frac{1}{(1 - x^2)^2} = \frac{1}{(1 + x)^2}.$$

Consequently,

$$\lim_{x \to 1-}(1 - x)\sum_k a_k x^k = \lim_{x \to 1-}\frac{1}{(1 + x)^2} = \frac{1}{4};$$

that is, the given sequence $\{a_n\}$ is Abel-summable to $1/4$, proving our claim.

2.4 Excercise

Exercise 2.1   Prove that the following series is $(C, 1)$ summable to 0:
a) $1 - 1 - 1 + 1 + 1 - 1 - 1 + 1 + 1 - 1 - 1 + 1 + \cdots$;
b) $\frac{1}{2} - 1 + \frac{1}{2} + \frac{1}{2} - 1 + \frac{1}{2} + \frac{1}{2} - 1 + \frac{1}{2} + \cdots$.

Exercise 2.2   Is the series $1 - 2 + 3 - 4 + \cdots$ $(C, 1)$ summable? Justify your answer.

Exercise 2.3   Give a bounded sequence that is not Abel summable and justify your answer.

Exercise 2.4   Let $x_n = (1 + (-1)^n)/n$, $n = 1, 2, \cdots$. Prove that $\{x_n\}$ is $(C, 1)$ summable to $1/2$.

Exercise 2.5   Give an example of an unbounded sequence that is $(C, 1)$ summable and justify your answer.

Exercise 2.6   Give an example of a sequence of 0s and 1s that is not $(C, 1)$ summable. Why?

Exercise 2.7   If $\{x_n\}$ is $(C, 1)$ summable to s, then every subsequence of $\{x_n\}$ is $(C, 1)$ summable to s. Is this statement true? Justify your answer.

Exercise 2.8   Give an example of a divergent sequence of 0s and 1s that is (i) Abel summable and (ii) not Abel summable.

Exercise 2.9   Let $\{x_n\}$ be defined by

$$x_n = \begin{cases} 1, & \text{if } n = 2^k; \\ 0, & \text{otherwise.} \end{cases}$$

Is $\{x_n\}$ $(C, 1)$ summable? Why?

Exercise 2.10   Prove that the $(C, 1)$ method is a particular case of a Weighted Mean method.

Exercise 2.11   Give an example of a Nörlund method, which is (i) regular (other than Example 2.1) and (ii) not regular (other than Example 2.2).

Exercise 2.12   Give an example of a Weighted Mean method, which is (i) regular and (ii) not regular.

Exercise 2.13   For the given series $\sum_n a_n$, let

$$s_n = \sum_{k=1}^{n} a_k, \quad t_n = \sum_{k=1}^{n} k a_k, \quad \sigma_n = \frac{1}{n} \sum_{k=1}^{n} s_k, \quad n = 1, 2, \cdots.$$

Prove that
(a) $t_n = (n + 1)s_n - n\sigma_n$, $n = 1, 2, \cdots$;
(b) If $\sum_n a_n$ is $(C, 1)$ summable, then $\sum_n a_n$ converges if and only if $t_n = o(n)$, $n \to \infty$;
(c) $\sum_n a_n$ is $(C, 1)$ summable if and only if $\sum_n t_n / n(n + 1)$ converges.

Hint. Part (a) can be proved easily by substituting for $\sigma_n, s_k, k = 1, 2, \cdots, n$. For proving Part (b), let $\sum_n a_n$ be $(C, 1)$ summable to s. Then, $\sigma_n \to s, n \to \infty$. Now

$$t_n / n = (1 + 1/n)s_n - \sigma_n,$$

by part (a). So, $\lim_n t_n / n = 0$ if and only if $\lim_n s_n = s$. In other words, $\sum_n a_n$ converges to s if and only if $t_n = o(n)$, $n \to \infty$. Part (c) can be similarly proved.

References

1 Hardy, G.H.: Divergent Series. Oxford University Press, Oxford (1949).
2 Powell, R.E. and Shah, S.M.: Summability Theory and Applications. Prentice-Hall of India, Delhi (1988).

3

Special Summability Methods II

For special methods of summability, standard references are [1, 3].

3.1 The Natarajan Method and the Abel Method

In an attempt to generalize the Nörlund method, Natarajan introduced the (M, λ_n) method (see [2]) as follows:

Definition 3.1 Given a sequence $\{\lambda_n\}$ of numbers such that $\sum_n |\lambda_n| < \infty$, the (M, λ_n) method is defined by the infinite matrix (a_{nk}), where

$$a_{nk} = \begin{cases} \lambda_{n-k}, & k \leq n; \\ 0, & k > n. \end{cases}$$

Remark 3.1 The (M, λ_n) method reduces to the well-known Y-method, when $\lambda_0 = \lambda_1 = 1/2$ and $\lambda_n = 0, n \geq 2$.

Note that the Natarajan method (M, λ_n) is a nontrivial summability method, that is, it is not equivalent to convergence. Consider the following example.

Example 3.1 Take any (M, λ_n) method. Then, we have $\sum_n |\lambda_n| < \infty$. Consider the sequence $\{1, 0, 1, 0, \dots \}$, which is not convergent. If $\{\sigma_n\}$ is the (M, λ_n)-transform of $\{1, 0, 1, 0, \dots \}$, then

$$\sigma_n = \lambda_0 + \lambda_2 + \lambda_4 + \cdots + \lambda_{2k}, \text{ if } n = 2k \text{ or } 2k + 1.$$

Now

$$\sum_k |\lambda_{2k}| \leq \sum_n |\lambda_n| < \infty,$$

so that $\{\sigma_n\}$ converges to s (say). Thus, $\{1, 0, 1, 0, \dots \}$ is (M, λ_n)-summable to s. Similarly, the series $1 - 1 + 1 - 1 + \cdots$, whose partial sum sequence is

An Introductory Course in Summability Theory, First Edition. Ants Aasma, Hemen Dutta, and P.N. Natarajan.
© 2017 John Wiley & Sons, Inc. Published 2017 by John Wiley & Sons, Inc.

$\{1, 0, 1, 0, \ldots\}$, is (M, λ_n)-summable. In particular, if $\lambda_0 = \lambda_1 = 1/2$, $\lambda_n = 0$, $n \geq 2$, the (M, λ_n) method reduces to the Y-method. The nonconvergent sequence $\{1, 0, 1, 0, \ldots\}$ and the nonconvergent series $1 - 1 + 1 - 1 + \cdots$ are Y-summable to $1/2$. The reader should try other examples as well.

Theorem 3.1 The (M, λ_n)-method is regular if and only if

$$\sum_n \lambda_n = 1. \tag{3.1}$$

Proof: Since $\Sigma_n |\lambda_n| < \infty$, then

$$\sup_{n \geq 0} \sum_k |a_{nk}| = \sup_{n \geq 0} \sum_{k=0}^{n} |a_{nk}| = \sup_{n \geq 0} \sum_{k=0}^{n} |\lambda_{n-k}| = \sup_{n \geq 0} \sum_{k=0}^{n} |\lambda_k| < \infty,$$

and

$$\lim_{n \to \infty} a_{nk} = \lim_{n \to \infty} \lambda_{n-k} = 0,$$

since $\lim_{n \to \infty} \lambda_n = 0$. Thus, $(M, \lambda_n) \equiv (a_{nk})$ is regular if and only if

$$1 = \lim_{n \to \infty} \sum_k a_{nk} = \lim_{n \to \infty} \sum_{k=0}^{n} a_{nk} = \lim_{n \to \infty} \sum_{k=0}^{n} \lambda_{n-k} = \lim_{n \to \infty} \sum_{k=0}^{n} \lambda_k = \sum_k \lambda_k,$$

completing the proof.

We now have the following theorem.

Theorem 3.2 Any two regular methods, (M, λ_n) and (M, μ_n), are consistent.

Proof: Let (M, λ_n) and (M, μ_n) be two regular methods. Let $\{u_n\}$, $\{v_n\}$ be the (M, λ_n), (M, μ_n)-transforms of a sequence $\{x_n\}$, respectively. That is,

$$u_n = \lambda_0 x_n + \lambda_1 x_{n-1} + \cdots + \lambda_n x_0,$$

$$v_n = \mu_0 x_n + \mu_1 x_{n-1} + \cdots + \mu_n x_0,$$

for all n. Let $\lim_{n \to \infty} u_n = s$, $\lim_{n \to \infty} v_n = t$. We claim that $s = t$. Let

$$\gamma_n = \lambda_0 \mu_n + \lambda_1 \mu_{n-1} + \cdots + \lambda_n \mu_0,$$

for all n. Now

$$w_n = \gamma_0 x_n + \gamma_1 x_{n-1} + \cdots + \gamma_n x_0,$$

$$= (\lambda_0 \mu_0) x_n + (\lambda_0 \mu_1 + \lambda_1 \mu_0) x_{n-1}$$

$$+ \cdots + (\lambda_0 \mu_n + \lambda_1 \mu_{n-1} + \cdots + \lambda_n \mu_0) x_0$$

$$= \lambda_0(\mu_0 x_n + \mu_1 x_{n-1} + \cdots + \mu_n x_0)$$
$$+ \lambda_1(\mu_0 x_{n-1} + \mu_1 x_{n-2} + \cdots + \mu_{n-1} x_0)$$
$$+ \cdots + \lambda_n(\mu_0 x_0)$$
$$= \lambda_0 v_n + \lambda_1 v_{n-1} + \cdots + \lambda_n v_0.$$

Thus, $\{w_n\}$ is the (M, λ_n)-transform of the sequence $\{v_k\}$. Since $\lim_{k \to \infty} v_k = t$ and (M, λ_n) is regular, it follows that

$$\lim_{n \to \infty} w_n = t.$$

In a similar manner, we can prove that

$$\lim_{n \to \infty} w_n = s,$$

so that $s = t$, completing the proof.

Definition 3.2 Let $s = \{s_0, s_1, s_2, \ldots\}$, $\bar{s} = \{0, s_0, s_1, s_2, \ldots\}$ and $s^* = \{s_1, s_2, \ldots\}$. The summability method A is said to be "translative" if $\bar{s}, s^*$ are A-summable to t, whenever, s is A-summable to t.

The following result exhibits plenty of translative methods.

Theorem 3.3 Every (M, λ_n) method is translative.

Proof: Writing $A \equiv (M, \lambda_n)$, we have

$$(A\bar{s})_n = \lambda_n 0 + \lambda_{n-1} s_0 + \lambda_{n-2} s_1 + \cdots + \lambda_0 s_{n-1}$$
$$= \lambda_{n-1} s_0 + \lambda_{n-2} s_1 + \cdots + \lambda_0 s_{n-1} = u_{n-1},$$

where

$$u_n = \lambda_n s_0 + \lambda_{n-1} s_1 + \cdots + \lambda_0 s_n,$$

for all n. So, if $u_n \to t, n \to \infty$, then

$$(A\bar{s})_n \to t, n \to \infty.$$

Also, as $n \to \infty$, we obtain

$$(As^*)_n = \lambda_n s_1 + \lambda_{n-1} s_2 + \cdots + \lambda_0 s_{n+1}$$
$$= (\lambda_{n+1} s_0 + \lambda_n s_1 + \cdots + \lambda_0 s_{n+1}) - \lambda_{n+1} s_0 = u_{n+1} - \lambda_{n+1} s_0 \to t,$$

since $u_n \to t, n \to \infty$ and $\lambda_n \to 0, n \to \infty$. Thus, (M, λ_n) is translative.

We now prove an inclusion theorem for (M, λ_n) methods.

Theorem 3.4 (Inclusion theorem)   Given the methods (M, λ_n), (M, μ_n),

$$(M, \lambda_n) \subseteq (M, \mu_n)$$

if and only if

$$\sum_n |k_n| < \infty \text{ and } \sum_n k_n = 1, \tag{3.2}$$

where

$$\frac{\mu(x)}{\lambda(x)} = k(x) = \sum_n k_n x^n,$$

$$\lambda(x) = \sum_n \lambda_n x^n,$$

$$\mu(x) = \sum_n \mu_n x^n.$$

Proof: As in Hardy [1, pp. 65–68], let

$$\lambda(x) = \sum_n \lambda_n x^n, \mu(x) = \sum_n \mu_n x^n.$$

Both of the series on the right converge if $|x| < 1$. Let $\{u_n\}$, $\{v_n\}$ be the (M, λ_n), (M, μ_n)-transforms of $\{s_n\}$, respectively. If $|x| < 1$, then

$$\sum_n v_n x^n = \sum_n (\mu_0 s_n + \mu_1 s_{n-1} + \cdots + \mu_n s_0) x^n$$

$$= \left(\sum_n \mu_n x^n \right) \left(\sum_n s_n x^n \right)$$

$$= \mu(x) s(x).$$

Similarly,

$$\sum_n u_n x^n = \lambda(x) s(x), \text{ if } |x| < 1.$$

Now

$$k(x) \lambda(x) = \mu(x),$$

$$k(x) \lambda(x) s(x) = \mu(x) s(x);$$

that is,

$$k(x) \left(\sum_n u_n x^n \right) = \left(\sum_n v_n x^n \right).$$

Thus,

$$v_n = k_0 u_n + k_1 u_{n-1} + \cdots + k_n u_0 = \sum_j a_{nj} u_j,$$

where

$$a_{nj} = \begin{cases} k_{n-j}, & j \leq n; \\ 0, & k > n. \end{cases}$$

If $(M, \lambda_n) \subseteq (M, \mu_n)$, then the infinite matrix (a_{nj}) is regular. So, appealing to Theorem 1.1, we get

$$\sup_{n \geq 0} \sum_j |a_{nj}| < \infty, \text{ or } \sup_{n \geq 0} \sum_{j=0}^n |a_{nj}| < \infty;$$

that is,

$$\sup_{n \geq 0} \sum_{j=0}^n |k_{n-j}| < \infty, \text{ or } \sup_{n \geq 0} \sum_{j=0}^n |k_j| < \infty;$$

that is,

$$\sum_j |k_j| < \infty.$$

Also, $\lim_{n \to \infty} \Sigma_j a_{nj} = 1$ implies that $\Sigma_n k_n = 1$.

Conversely, if $\Sigma_n |k_n| < \infty$ and $\Sigma_n k_n = 1$, then it follows that (a_{nj}) is regular and so $\lim_{j \to \infty} u_j = t$ implies that $\lim_{n \to \infty} v_n = t$. Thus, $(M, \lambda_n) \subseteq (M, \mu_n)$.

As a consequence of Theorem 3.4, we have the following result.

Theorem 3.5 (Equivalence theorem) The methods (M, λ_n), (M, μ_n) are equivalent; that is,

$$(M, \lambda_n) \subseteq (M, \mu_n), \text{ and conversely,}$$

if and only if

$$\sum_n |k_n| < \infty, \sum_n |h_n| < \infty,$$

and

$$\sum_n k_n = 1, \sum_n h_n = 1,$$

where

$$\frac{\mu(x)}{\lambda(x)} = k(x) = \sum_n k_n x^n,$$

$$\frac{\lambda(x)}{\mu(x)} = h(x) = \sum_n h_n x^n,$$

$$\lambda(x) = \sum_n \lambda_n x^n, \mu(x) = \sum_n \mu_n x^n.$$

The following theorem gives the connection between the Natarajan method (M, λ_n) and the Abel method (see [2]).

Theorem 3.6 [2, Theorem 4.2] If $\{a_n\}$ is (M, λ_n)-summable to s, where (M, λ_n) is regular, then $\{a_n\}$ is Abel-summable to s.

Proof: Let $\{u_n\}$ be the (M, λ_n)-transform of the sequence $\{a_k\}$ so that

$$u_n = \lambda_0 a_n + \lambda_1 a_{n-1} + \cdots + \lambda_n a_0,$$

for all n. Then $\lim_{n \to \infty} u_n = s$. Now

$$\left(\sum_n \lambda_n x^n \right) \left(\sum_n a_n x^n \right) = \sum_n u_n x^n,$$

$$\left[(1-x) \left(\sum_n a_n x^n \right) \right] \left(\sum_n \lambda_n x^n \right) = (1-x) \left(\sum_n u_n x^n \right),$$

$$\left[(1-x) \left(\sum_n a_n x^n \right) \right] (1-x) \left(\sum_n \lambda_n x^n \right) \left(\sum_n x^n \right) = (1-x) \left(\sum_n u_n x^n \right);$$

that is,

$$\left[(1-x) \left(\sum_n a_n x^n \right) \right] (1-x) \left(\sum_n \Lambda_n x^n \right) = (1-x) \left(\sum_n u_n x^n \right), \quad (3.3)$$

where $\Lambda_n = \sum_{k=0}^{n} \lambda_k$, for all n. Taking the limit as $x \to 1-$ in (3.3), we have

$$\lim_{x \to 1-} (1-x) \left(\sum_n a_n x^n \right) = \lim_{x \to 1-} (1-x) \left(\sum_n u_n x^n \right), \quad (3.4)$$

noting that

$$\lim_{x \to 1-} (1-x) \left(\sum_n \Lambda_n x^n \right) = 1$$

since

$$\lim_{n \to \infty} \Lambda_n = \sum_n \lambda_n = 1,$$

and the Abel method is regular. Since $\lim_{n \to \infty} u_n = s$,

$$\lim_{x \to 1-} (1-x) \left(\sum_n u_n x^n \right) = s,$$

again using the fact that the Abel method is regular.

It now follows from (3.4) that

$$\lim_{x \to 1-} (1 - x) \sum_n a_n x^n = s;$$

that is, $\{a_n\}$ is Abel-summable to s.

3.2 The Euler and Borel Methods

The Euler summability method is defined as follows:

Definition 3.3 Let $r \in \mathbb{C} \setminus \{1, 0\}$, $\mathbb{C}$ being the field of complex numbers. The Euler method of order r or the (E, r) method is defined by the infinite matrix $(e_{nk}^{(r)})$, where

$$e_{nk}^{(r)} = \begin{cases} {}^n c_k r^k (1 - r)^{n-k}, & k \leq n; \\ 0, & k > n. \end{cases}$$

For $r \in \{1, 0\}$, the (E, r) method is defined respectively by the infinite matrices $(e_{nk}^{(1)})$ and $(e_{nk}^{(0)})$, where

$$e_{nk}^{(1)} = \begin{cases} 1, & k = n; \\ 0, & k \neq n. \end{cases}$$

$$e_{nk}^{(0)} = 0, n = 0, 1, 2, \dots; \quad k = 1, 2, \dots;$$

$$e_{n0}^{(0)} = 1, n = 0, 1, 2, \dots.$$

We now prove a criterion for the regularity of the (E, r) method.

Theorem 3.7 The (E, r) method is regular if and only if r is real and $0 < r \leq 1$.

Proof: We first note that the series $\sum_{n=k}^{\infty} {}^n c_k w^{n-k}$ converges for $|w| < 1$. For every fixed k,

$$\lim_{n \to \infty} e_{nk}^{(r)} = \lim_{n \to \infty} {}^n c_k r^k (1 - r)^{n-k}$$

$$= r^k \lim_{n \to \infty} {}^n c_k (1 - r)^{n-k}.$$

If $|1 - r| \geq 1$, then $\lim_{n \to \infty} {}^n c_k (1 - r)^{n-k} \neq 0$. However, if $|1 - r| < 1$, then

$$\sum_{n=k}^{\infty} {}^n c_k (1 - r)^{n-k}$$

converges, and so

$$\lim_{n \to \infty} {}^n c_k (1 - r)^{n-k} = 0.$$

As a consequence, we have

$$\lim_{n \to \infty} e_{nk}^{(r)} = 0 \tag{3.5}$$

if and only if $|1 - r| < 1$.

Again,

$$\sum_k e_{nk}^{(r)} = \sum_{k=0}^{n} e_{nk}^{(r)} = \sum_{k=0}^{n} {}^n c_k r^k (1-r)^{n-k} = \{r + (1-r)\}^n = 1,$$

so that

$$\lim_{n \to \infty} \sum_k e_{nk}^{(r)} = 1.$$

Finally,

$$\sum_k |e_{nk}^{(r)}| = \sum_{k=0}^{n} |e_{nk}^{(r)}| = \sum_{k=0}^{n} {}^n c_k |r|^k |1-r|^{n-k} = (|r| + |1-r|)^n,$$

and so,

$$\sup_n \sum_k |e_{nk}^{(r)}| \le M < \infty, M > 0,$$

if and only if

$$|r| + |1 - r| \le 1.$$

By the triangle inequality, we know that

$$|r| + |1 - r| \ge |r + (1 - r)| = 1$$

for all $r \in \mathbb{C}$. Thus

$$|r| + |1 - r| \le 1$$

if and only if

$$|r| + |1 - r| = 1;$$

that is, r is real and $0 \le r \le 1$.

With the restriction $|1 - r| < 1$, $r \ne 0$, the matrix $(e_{nk}^{(r)})$ is regular if and only if r is real and $0 < r \le 1$, completing the proof.

Theorem 3.8 *If $rs \ne 0$, the product $(e_{nk}^{(r)}) (e_{nk}^{(s)})$ is the matrix of the (E, rs) method; that is,*

$$(e_{nk}^{(r)})(e_{nk}^{(s)}) = (e_{nk}^{(rs)}). \tag{3.6}$$

Proof: Let $(e_{nk}^{(r)})(e_{nk}^{(s)}) = (a_{nj})$. Note that $a_{nj} = 0, j > n$. Since $(E, 1)$ is ordinary convergence, the result is trivial when $r = 1$ or $s = 1$. Let $r \neq 1, s \neq 1$ and $j \leq n$. Then

$$a_{nj} = \sum_{k=j}^{n} e_{nk}^{(r)} e_{nk}^{(s)}.$$

If $j = n$, then $a_{nn} = e_{nn}^{(r)} e_{nn}^{(s)} = r^n s^n = (rs)^n$. If $j < n$, then we can write

$$a_{nj} = \sum_{k=j}^{n} e_{nk}^{(r)} e_{kj}^{(s)}$$

$$= \sum_{k=j}^{n} {}^{n}c_k r^k (1 - r)^{n-k} \quad {}^{k}c_j s^j (1 - s)^{k-j}$$

$$= s^j \frac{(1 - r)^n}{(1 - s)^j} \sum_{k=j}^{n} {}^{n}c_k {}^{k}c_j r^k \left(\frac{1 - s}{1 - r}\right)^k$$

$$= {}^{n}c_j (rs)^j (1 - r)^{n-j} \sum_{k=j}^{n} {}^{(n-j)}c_{(k-j)} \left\{\frac{r(1 - s)}{1 - r}\right\}^{k-j}$$

$$= {}^{n}c_j (rs)^j (1 - r)^{n-j} \left[1 + \frac{r(1 - s)}{1 - r}\right]^{n-j}$$

$$= {}^{n}c_j (rs)^j (1 - r)^{n-j} \left[\frac{1 - r + r - rs}{1 - r}\right]^{n-j}$$

$$= {}^{n}c_j (rs)^j (1 - rs)^{n-j},$$

completing the proof.

The following important result is an immediate consequence of Theorem 3.8.

Corollary 3.1   If $r \neq 0$, then (E, r) is invertible and $(E, r)^{-1} = (E, 1/r)$.

Theorem 3.9   (Inclusion theorem)   If $0 < |s| \leq |r|$ and $|s| + |r - s| = |r|$, then

$$(E, r) \subseteq (E, s).$$

Proof: Let $0 < |s| \leq |r|$ and $|s| + |r - s| = |r|$. Let $\{x_k\}$ be (E, r)-summable y (say); that is,

$$\lim_{n \to \infty} t_n = y,$$

where

$$t_n = \sum_k e^{(r)}_{nk} x_k = \sum_{k=0}^{n} e^{(r)}_{nk} x_k.$$

Using Corollary 3.1, we obtain

$$\sum_{n=0}^{j} e^{\left(\frac{1}{r}\right)}_{jn} t_n = \sum_{n=0}^{j} e^{\left(\frac{1}{r}\right)}_{jn} \left(\sum_{k=0}^{n} e^{(r)}_{nk} x_k \right) = \sum_{k=0}^{j} \left(\sum_{n=k}^{j} e^{\left(\frac{1}{r}\right)}_{jn} e^{(r)}_{nk} \right) x_k = x_j. \qquad (3.7)$$

Now, using (3.7) and (3.6), we obtain

$$\sigma_n = \sum_j e^{(s)}_{kj} x_j = \sum_{j=0}^{k} e^{(s)}_{kj} x_j = \sum_{j=0}^{k} e^{(s)}_{kj} \left(\sum_{n=0}^{j} e^{\left(\frac{1}{r}\right)}_{jn} t_n \right),$$

$$= \sum_{n=0}^{k} \left(\sum_{j=n}^{k} e^{(s)}_{kj} e^{\left(\frac{1}{r}\right)}_{jn} \right) t_n = \sum_{n=0}^{k} e^{\left(\frac{s}{r}\right)}_{kn} t_n.$$

Since $0 < |s| \le |r|$ and $|s| + |r - s| = |r|$, we have $0 < s/r \le 1$, so that $(E, s/r)$ is regular, using Theorem 3.7. Consequently, $\lim_{n \to \infty} t_n = y$ implies that $\lim_{n \to \infty} \sigma_n = y$; that is, $(E, r) \subseteq (E, s)$.

Definition 3.4 A sequence $\{z_k\}$ is said to be Borel summable or (B)-summable to y if

$$\sum_k z_k \frac{x^k}{k!}$$

converges for all real x and

$$\lim_{x \to \infty} e^{-x} \sum_k z_k \frac{x^k}{k!}$$

exists finitely and is equal to y.
The above method is also known as the "Borel exponential method."

Theorem 3.10 The Borel exponential method is regular.

Proof: Let $\{z_k\}$ converge to z (say). The

$$\overline{\lim_{n \to \infty}} \left| \frac{z_n}{n!} \right|^{1/n} = 0,$$

and hence the series $\Sigma_k z_k x^k / k!$ converges for all real x. It is easy to see that

$$\left| e^{-x} \sum_k z_k \frac{x^k}{k!} - z \right| = \left| e^{-x} \sum_k z_k \frac{x^k}{k!} - e^{-x} \sum_k \frac{x^k}{k!} z \right| = \left| e^{-x} \sum_k (z_k - z) \frac{x^k}{k!} \right|.$$

For each $\epsilon > 0$, a positive integer K exists such that

$$|z_k - z| < \frac{\epsilon}{2}, k \geq K.$$

Since $\{z_k - z\}$ converges, there exists an $M > 0$, such that

$$|z_k - z| \leq M.$$

Thus, for $x \geq 1$, we can write

$$\left| e^{-x} \sum_k z_k \frac{x^k}{k!} - z \right|$$

$$= \left| e^{-x} \left\{ \sum_{k=0}^{K-1} (z_k - z) \frac{x^k}{k!} + \sum_{k=K}^{\infty} (z_k - z) \frac{x^k}{k!} \right\} \right|$$

$$\leq \left| e^{-x} \sum_{k=0}^{K-1} (z_k - z) \frac{x^k}{k!} \right| + \left| e^{-x} \sum_{k=K}^{\infty} (z_k - z) \frac{x^k}{k!} \right|$$

$$< e^{-x} K M x^{K-1} + \left(e^{-x} \sum_k \frac{x^k}{k!} \right) \frac{\epsilon}{2}.$$

Note that

$$\frac{x^{K-1}}{e^x} \to 0, x \to \infty.$$

Now choose $\delta > 1$, such that

$$\frac{x^{K-1}}{e^x} < \frac{\epsilon}{2MK}.$$

Thus, for $x > \delta$, we have

$$\left| e^{-x} \sum_k z_k \frac{x^k}{k!} - z \right| < MK \frac{\epsilon}{2MK} + \frac{\epsilon}{2} = \epsilon,$$

that is,

$$\lim_{x \to \infty} e^{-x} \sum_k z_k \frac{x^k}{k!} = z,$$

and the Borel exponential method is regular.

Often the Borel exponential method is replaced by a matrix method, called the Borel matrix method of summability. It is defined by the infinite matrix (a_{nk}), where

$$a_{nk} = e^{-n} \frac{n^k}{k!},$$

for all n, k.

The following result gives the connection between the Euler and Borel methods.

Theorem 3.11   If $r > 0$, then $(E, r) \subseteq (B)$.

Proof: Let $r > 0$ and let the sequence $\{z_k\}$ be (E, r)-summable to y; that is,

$$\lim_{n \to \infty} t_n = y,$$

where

$$t_n = \sum_k e_{nk}^{(r)} z_k = \sum_{k=0}^{n} e_{nk}^{(r)} z_k.$$

Using Corollary 3.1, we get

$$\sum_{n=0}^{j} e_{jn}^{\left(\frac{1}{r}\right)} t_n = \sum_{n=0}^{j} e_{jn}^{\left(\frac{1}{r}\right)} \left(\sum_{k=0}^{n} e_{nk}^{(r)} z_k \right) = \sum_{k=0}^{j} \left(\sum_{n=k}^{j} e_{jn}^{\left(\frac{1}{r}\right)} e_{nk}^{(r)} \right) z_k = z_j.$$

Again,

$$\sum_j z_j \frac{x^j}{j!} = \sum_j \left(\sum_{n=0}^{j} e_{jn}^{\left(\frac{1}{r}\right)} t_n \right) \frac{x^j}{j!} = \sum_n \left(\sum_{j=n}^{\infty} e_{jn}^{\left(\frac{1}{r}\right)} \frac{x^j}{j!} \right) t_n$$

$$= \sum_n \left[\sum_{j=n}^{\infty} {}^j c_n \left(\frac{1}{r} \right)^n \left(1 - \frac{1}{r} \right)^{j-n} \frac{x^j}{j!} \right] t_n$$

$$= \sum_n \frac{1}{n!} \left(\frac{x}{r} \right)^n t_n \left[\sum_{j=n}^{\infty} \frac{\left\{ x \left(1 - \frac{1}{r} \right) \right\}^{j-n}}{(j-n)!} \right]$$

$$= e^{\left(1 - \frac{1}{r}\right)x} \left[\sum_n \frac{1}{n!} \left(\frac{x}{r} \right)^n t_n \right].$$

We note that interchanging the order of summation is allowable, since the double series in these relations converges absolutely. So,

$$e^{-x} \sum_j z_j \frac{x^j}{j!} = e^{-\frac{x}{r}} \left[\sum_n \frac{1}{n!} \left(\frac{x}{r} \right)^n t_n \right],$$

$$\lim_{x \to \infty} e^{-x} \sum_j z_j \frac{x^j}{j!} = \lim_{x \to \infty} e^{-\frac{x}{r}} \left[\sum_n \frac{1}{n!} \left(\frac{x}{r} \right)^n t_n \right] = y,$$

since the Borel method (B) is regular and $\lim_{n \to \infty} t_n = y$. Hence,

$$(E, r) \subseteq (B).$$

3.3 The Taylor Method

The Taylor method, also known as the circle method, was introduced by G.H. Hardy and J.E. Littlewood in 1916.

Definition 3.5 Let $r \in \mathbb{C} \setminus \{0\}$. The Taylor method of order r, denoted by (T, r), is defined by the infinite matrix $(t_{nk}^{(r)})$, where

$$t_{nk}^{(r)} = \begin{cases} 0, & \text{if } k < n; \\ {}^k c_n r^{k-n}(1-r)^{n+1}, & \text{if } k \geq n. \end{cases}$$

For $r = 0$, the Taylor method is defined by the infinite matrix $t_{nk}^{(0)}$ where $r = 0$,

$$t_{nk}^{(0)} = \begin{cases} 1, & \text{if } k = n; \\ 0, & \text{if } k \neq n. \end{cases}$$

Theorem 3.12 The Taylor method (T, r) is regular if and only if r is real and $0 \leq r < 1$.

Proof: If $r = 1$, then $t_{nk}^{(1)} = 0$ and so the $(T, 1)$ matrix transforms all sequences into the sequence $\{0, 0, \ldots, 0, \ldots \}$. So, let $r \neq 1$. Let k be fixed. Since $t_{nk}^{(r)} = 0$, $n > k$, it follows that

$$\lim_{n \to \infty} t_{nk}^{(r)} = 0, \text{ for all } k.$$

Then

$$\sum_k t_{nk}^{(r)} = \sum_{k=n}^{\infty} t_{nk}^{(r)}$$

$$= \sum_{k=n}^{\infty} {}^k c_n r^{k-n}(1-r)^{n+1}$$

$$= (1-r)^{n+1} \sum_{k=n}^{\infty} {}^k c_n r^{k-n},$$

where the series converges only if $|r| < 1$ and, in this case,

$$(1-r)^{n+1} \sum_{k=n}^{\infty} {}^k c_n r^{k-n} = (1-r)^{n+1}(1-r)^{-(n+1)} = 1;$$

that is,

$$\sum_k t_{nk}^{(r)} = 1,$$

so that

$$\lim_{n \to \infty} \sum_k t_{nk}^{(r)} = 1$$

if and only if

$$|r| < 1.$$

Finally,

$$\sum_k |t_{nk}^{(r)}| = \sum_{k=n}^{\infty} |t_{nk}^{(r)}| = \sum_{k=n}^{\infty} {}^k c_n |r|^{k-n} |1 - r|^{n+1}$$

$$= |1 - r|^{n+1} \sum_{k=n}^{\infty} {}^k c_n |r|^{k-n} = \frac{|1 - r|^{n+1}}{(1 - |r|)^{n+1}},$$

since $|r| < 1$. Hence,

$$\sup_{n \geq 0} \left(\sum_k |t_{nk}^{(r)}| \right) < \infty$$

if and only if

$$\frac{|1 - r|}{1 - |r|} \leq 1 \text{ or } |1 - r| \leq 1 - |r|.$$

By the triangle inequality, we have

$$|1 - r| \geq 1 - |r|,$$

so

$$\sup_{n \geq 0} \left(\sum_k |t_{nk}^{(r)}| \right) < \infty$$

if and only if

$$|1 - r| = 1 - |r|;$$

i.e., r is real and $0 \leq r \leq 1$. Thus, (T, r) is regular if and only if

$$|r| < 1 \text{ and } 0 \leq r \leq 1;$$

i.e., $0 \leq r < 1$.

Theorem 3.13   The product of the (T, r) and the (T, s) matrices is the matrix

$$(1 - r)(1 - s)(E, (1 - r)(1 - s))',$$

where $(a_{nk})'$ denotes the "transpose" of the matrix (a_{nk}).

Proof: Let $(\alpha_{nk}) = (t_{nj}^{(r)})(t_{jk}^{(s)})$, that is, $\alpha_{nk} = 0$, if $k < n$ and, if $k \geq n$, then

$$\alpha_{nk} = \sum_{j=n}^{k} t_{nj}^{(r)} t_{jk}^{(s)}$$

$$= \sum_{j=n}^{k} {}^{j}c_{n} r^{j-n} (1-r)^{n+1} {}^{k}c_{j} s^{k-j} (1-s)^{j+1}$$

$$= {}^{k}c_{n}\{(1-r)(1-s)\}^{n+1} s^{k-n} \sum_{j=n}^{k} {}^{(k-n)}c_{(j-n)} \left\{ \frac{r(1-s)}{s} \right\}^{j-n}$$

$$= {}^{k}c_{n}\{(1-r)(1-s)\}^{n+1} s^{k-n} \left[1 + \frac{r(1-s)}{s} \right]^{k-n}$$

$$= {}^{k}c_{n}\{(1-r)(1-s)\}^{n+1} (r+s-rs)^{k-n}.$$

Now, for $u = (1-r)(1-s)$ and $k \geq n$, we obtain, by Definition 3.3,

$$e_{kn}^{(u)} = {}^{k}c_{n}\{(1-r)(1-s)\}^{n}\{1 - (1-r)(1-s)\}^{k-n}$$

$$= {}^{k}c_{n}\{(1-r)(1-s)\}^{n}(r+s-rs)^{k-n}.$$

Thus, if $k \geq n$, then

$$\alpha_{nk} = (1-r)(1-s)e_{kn}^{(u)}.$$

Hence,

$$(T, r)(T, s) = (1-r)(1-s)(E, (1-r)(1-s))',$$

where (T, r) denotes the matrix of the (T, r) method, and so on. This completes the proof.

The following is an immediate and important consequence of Theorem 3.13.

Theorem 3.14   If $r \neq 1$, then (T, r) is invertible and

$$(T, r)^{-1} = \left(T, \frac{-r}{1-r} \right).$$

Proof: Put $s = -\frac{r}{1-r}$ in Theorem 3.13. Then

$$T(r)T(s) = (1-r)(1-s)(E, (1-r)(1-s))'$$

$$= (1-r)\left(1 + \frac{r}{1-r} \right)\left(E, (1-r)\left(1 + \frac{r}{1-r} \right) \right)'$$

$$= (1-r)\frac{1}{1-r}\left(E, (1-r)\left(\frac{1}{1-r} \right) \right)' = (E, 1)' = (E, 1),$$

where $(E, 1)$ stands for usual convergence. This shows that (T, r) is invertible and

$$(T, r)^{-1} = \left(T, \frac{-r}{1-r} \right),$$

completing the proof.

3.4 The Hölder and Cesàro Methods

The Hölder method is defined as follows:

Definition 3.6 The $(H, 1)$ method is defined by the infinite matrix $(h_{nk}^{(1)})$, where,

$$h_{nk}^{(1)} = \begin{cases} \frac{1}{n+1}, & k \leq n; \\ 0, & k > n. \end{cases}$$

If m is a positive integer, the Hölder method of order m, denoted by (H, m), is defined by the infinite matrix $(h_{nk}^{(m)})$, with

$$(h_{nk}^{(m)}) = (h_{nk}^{(1)})(h_{nk}^{(m-1)}),$$

where the product of the two matrices denotes usual matrix multiplication.

Note that the $(H, 1)$ method is the $(C, 1)$ method, already introduced in Section 2.3.

Definition 3.7 Let $\{z_k\}$ be a sequence of (complex) numbers. Let α be any real number excluding the negative integers. Define the sequences $\{A_n^{(\alpha)}\}_{n=0}^{\infty}$ and $\{S_n^{(\alpha)}\}_{n=0}^{\infty}$ by

$$\sum_k A_k^{(\alpha)} x^k = (1-x)^{-\alpha-1}$$

and

$$\sum_k S_k^{(\alpha)} x^k = (1-x)^{-\alpha} \sum_k z_k x^k.$$

We say that the sequence $\{z_k\}$ is (C, α)-summable to y if

$$\lim_{k \to \infty} \frac{S_k^{(\alpha)}}{A_k^{(\alpha)}} = y.$$

We call (C, α) the Cesàro method of order α.

Let us now examine, in detail, the case when $\alpha = m$, where m is a positive integer. By the Binomial theorem, for $|x| < 1$, we have

$$(1 - x)^{-\alpha-1} = \sum_k {}^{(k+m)}c_m x^k.$$

Consequently, in this case,

$$A_k^{(m)} = {}^{(k+m)}c_m.$$

Also, we have

$$\sum_k S_k^{(m)} x^k = (1 - x)^{-m} \sum_k z_k x^k$$

$$= \left(\sum_j {}^{(j+m-1)}c_{m-1} x^j \right) \left(\sum_k z_k x^k \right) = \sum_k c_k x^k,$$

where

$$c_k = \sum_{t=0}^{k} {}^{(k-t+m-1)}c_{m-1} z_t.$$

Hence,

$$S_k^{(m)} = \sum_{t=0}^{k} {}^{(k-t+m-1)}c_{m-1} z_t.$$

So, (C, m) is defined by the infinite matrix $(c_{nk}^{(m)})$, where

$$c_{nk}^{(m)} = \begin{cases} \dfrac{{}^{(n-k+m-1)}c_{m-1}}{{}^{(n+m)}c_m}, & \text{if } k \leq n; \\ 0, & \text{if } k > n. \end{cases}$$

In particular, if $m = 1$, then

$$c_{nk}^{(1)} = \begin{cases} \dfrac{1}{n+1}, & \text{if } k \leq n; \\ 0, & \text{if } k > n. \end{cases}$$

In this case, we get the $(C, 1)$ method, which we introduced earlier.

We now take up the general case. Let α be any real number "excluding the negative integers." Now

$$(1 - x)^{-\alpha-1} = 1 + \sum_{k=1}^{\infty} \frac{(\alpha + 1)(\alpha + 2) \cdots (\alpha + k)}{k!} x^k,$$

so that

$$A_0^{(\alpha)} = 1, A_k^{(\alpha)} = \frac{(\alpha + 1)(\alpha + 2) \cdots (\alpha + k)}{k!}, \quad k = 1, 2, \dots.$$

Also,

$$\sum_k s_k^{(\alpha)} x^k = (1-x)^{-\alpha} \sum_k z_k x^k$$

$$= \left[1 + \sum_{k=1}^{\infty} \frac{\alpha(\alpha+1)\cdots(\alpha+k-1)}{k!} x^k \right] \left(\sum_j z_j x^j \right)$$

$$= \sum_k c_k^{(\alpha)} x^k,$$

where

$$c_k^{(\alpha)} = \sum_{j=0}^{k-1} \frac{\alpha(\alpha+1)\cdots(\alpha+k-j-1)}{(k-j)!} z_j + z_k.$$

Consequently, the (C, α) method is defined by the infinite matrix $(c_{kj}^{(\alpha)})$, where

$$c_{kj}^{(\alpha)} = \begin{cases} 1, & \text{if } k = j = 0; \\[2mm] \dfrac{1}{\frac{(\alpha+1)(\alpha+2)\cdots(\alpha+k)}{k!}}, & \text{if } k = j \neq 0; \\[4mm] \dfrac{\frac{\alpha(\alpha+1)\cdots(\alpha+k-j-1)}{(k-j)!}}{\frac{(\alpha+1)(\alpha+2)\cdots(\alpha+k)}{k!}}, & \text{if } k > j; \\[4mm] 0, & \text{otherwise.} \end{cases}$$

3.5 The Hausdorff Method

We conclude this chapter by introducing the Hausdorff method and study some of its properties. We need the following definition.

Definition 3.8   Let $x = \{x_k\}$ be a sequence of real numbers. Define

$$(\Delta^0 x)_n = x_n,$$
$$(\Delta^1 x)_n = x_n - x_{n+1},$$

and

$$(\Delta^j x)_n = (\Delta^{j-1} x)_n - (\Delta^{j-1} x)_{n+1}, j = 2, 3, \ldots .$$

The sequence $x = \{x_k\}$ is said to be "totally monotone" if

$$(\Delta^j x)_n \geq 0$$

for all n, j.

Example 3.2 The sequence $x = \{x_k\}$, where $x_k = r^k$, $0 \le r \le 1$, is totally monotone, since

$$(\Delta^j x)_n = r^n (1-r)^j.$$

Theorem 3.15 If $x = \{x_k\}$ is a sequence of real numbers, then

$$(\Delta^j x)_n = \sum_{k=0}^{j} {}^{j}c_k (-1)^k x_{n+k}.$$

Proof: The case $j = 0$ holds trivially. If $j = 1$, then

$$\sum_{k=0}^{1} {}^{1}c_k (-1)^k x_{n+k} = x_n - x_{n+1} = (\Delta^1 x)_n.$$

Let

$$(\Delta^j x)_n = \sum_{k=0}^{j} {}^{j}c_k (-1)^k x_{n+k}, j = 0, 1, 2, \ldots, m, m \ge 1.$$

Now

$$(\Delta^{m+1} x)_n = (\Delta^m x)_n - (\Delta^m x)_{n+1}$$

$$= \sum_{k=0}^{m} {}^{m}c_k (-1)^k x_{n+k} - \sum_{k=0}^{m} {}^{m}c_k (-1)^k x_{n+1+k}$$

$$= \sum_{k=0}^{m} {}^{m}c_k (-1)^k x_{n+k} - \sum_{k=1}^{m+1} {}^{m}c_{k-1} (-1)^{k-1} x_{n+k}$$

$$= x_n + \sum_{k=1}^{m} (-1)^k [{}^{m}c_k + {}^{m}c_{k-1}] x_{n+k} - (-1)^m x_{n+m+1}$$

$$= x_n + \sum_{k=1}^{m} (-1)^k \; {}^{m+1}c_k x_{n+k} - (-1)^m x_{n+m+1}$$

$$= \sum_{k=0}^{m+1} {}^{m+1}c_k (-1)^k x_{n+k},$$

completing the induction and proof of the theorem.

Definition 3.9 Define the matrix $\delta = (\delta_{nk})$ by

$$\delta_{nk} = \begin{cases} (-1)^k \; {}^{n}c_k, & \text{if } k \le n; \\ 0, & \text{if } k > n. \end{cases}$$

Note that $\delta^2 = \delta\delta$ is the identity matrix, that is, δ^2 is defined by the matrix (e_{nk}), where

$$e_{nk} = \begin{cases} 1, & \text{if } k = n; \\ 0, & \text{otherwise.} \end{cases}$$

Definition 3.10 The matrix $\mu = (\mu_{nk})$ is called a diagonal matrix if $\mu_{nk} = 0$ for all $n \neq k$. In this case, we write $\mu_{nn} = \mu_n$.

Definition 3.11 If $\mu = (\mu_{nk})$ is a diagonal matrix, then the method defined by the infinite matrix $u = (u_{nk})$, where

$$u = \delta\mu\delta = (\delta_{nm})(\mu_{mj})(\delta_{jk})$$

is called a Hausdorff method, denoted by (H, μ).

Note that $\delta\mu\delta$ is well defined since δ and μ are lower triangular matrices.

Theorem 3.16

(i) Any two Hausdorff methods commute; that is, if $(H, \mu) = (u_{nk})$ and $(H, \gamma) = (v_{nk})$ are two Hausdorff methods, then $(u_{nk})(v_{nk}) = (v_{nk})(u_{nk})$.

(ii) the product of any two Hausdorff methods is again a Hausdorff method.

Proof: Let $\delta\mu\delta = (u_{nk})$ and $\delta\gamma\delta = (v_{nk})$. Then, since the matrices are lower triangular, associativity of multiplication holds, so

$$\begin{aligned} (\delta\mu\delta)(\delta\gamma\delta) &= (\delta\mu)(\delta\delta)(\gamma\delta) \\ &= (\delta\mu)(\gamma\delta), \text{ since } \delta\delta \text{ is the identity matrix} \\ &= \delta(\mu\gamma)\delta \\ &= \delta(\gamma\mu)\delta \text{ since } \mu\gamma = \gamma\mu \\ &= (\delta\gamma)(\mu\delta) \\ &= (\delta\gamma)(\delta\delta)(\mu\delta), \text{ since } \delta\delta \text{ is the identity matrix} \\ &= (\delta\gamma\delta)(\delta\mu\delta), \end{aligned}$$

noting that $\mu\gamma = \gamma\mu$, since μ, γ are diagonal matrices. We also note that the product $(H, \mu)(H, \gamma)$ is the Hausdorff method $(H, \mu\gamma)$.

Corollary 3.2 A Hausdorff method (H, μ) is invertible if $\mu_n \neq 0$, for all n. In this case,

$$(H, \mu)^{-1} = (H, \gamma),$$

where $\gamma_n = 1/\mu_n$, for all n.

Example 3.3 Examples of Hausdorff methods are the $(C, 1)$ method ($\mu_n = 1/(n + 1)$, for all n) and the Euler method ($\mu_n = r^n$, for all n).

Theorem 3.17 Let (H, μ) be a Hausdorff method with $\mu_m \neq \mu_n$, $m \neq n$. If A is a lower triangular matrix that commutes with (H, μ), then A is a Hausdorff method.

Proof: Let $A = (a_{nk})$ be a lower triangular matrix, which commutes with (H, μ), that is, commutes with the matrix $\lambda = \delta \mu \delta$. Let $w = \delta A \delta$. So $\delta w \delta = \delta \delta A \delta \delta = A$, since $\delta \delta$ is the identity matrix. If w is a diagonal matrix, then A is a Hausdorff method. Now

$$w\mu = (\delta A \delta)(\delta \lambda \delta) = \delta A \delta \delta \lambda \delta = \delta A \lambda \delta$$

and

$$\mu w = (\delta \lambda \delta)(\delta A \delta) = \delta \lambda \delta \delta A \delta = \delta \lambda A \delta.$$

Since $A\lambda = \lambda A$, we have $w\mu = \mu w$. Let $w = (w_{nk})$. So

$$w\mu = (\ell_{nk}),$$

where

$$\ell_{nk} = \begin{cases} w_{nk}\mu_k, & \text{if } k \leq n; \\ 0, & \text{if } k > n, \end{cases}$$

and

$$\mu w = (r_{nk}),$$

where

$$r_{nk} = \begin{cases} \mu_n w_{nk}, & \text{if } k \leq n; \\ 0, & \text{if } k > n. \end{cases}$$

Thus, $\ell_{nk} = r_{nk}$ if and only if

$$w_{nk}\mu_k = \mu_n w_{nk} \text{ for all } k \leq n.$$

If $k = n$, it holds. If $k < n$, then

$$w_{nk}(\mu_k - \mu_n) = 0.$$

It now follows that

$$w_{nk} = 0,$$

since $\mu_k \neq \mu_n$ for $k \neq n$. Consequently, w is a diagonal matrix, completing the proof of the theorem.

Corollary 3.3 A lower triangular matrix A is a Hausdorff matrix if and only if it commutes with the $(C, 1)$ matrix.

Proof: The result follows using Theorem 3.17 noting that the $(C, 1)$ matrix is a Hausdorff matrix in which the diagonal elements differ.

Theorem 3.18 If $(H, \mu) = (h_{nk})$ is a Hausdorff method, then

$$h_{nk} = \begin{cases} {}^n c_k (\Delta^{n-k} \mu)_k, & \text{if } k \leq n; \\ 0, & \text{if } k > n. \end{cases}$$

Proof: We first note that

$$h_{nk} = \begin{cases} \displaystyle\sum_{j=k}^{n} {}^n c_j^{\,j} c_k (-1)^{j+k} \mu_j, & \text{if } k \leq n; \\ 0, & \text{if } k > n. \end{cases}$$

By Theorem 3.15, we obtain

$$\sum_{j=k}^{n} {}^n c_j^{\,j} c_k (-1)^{j+k} \mu_j = \sum_{j=k}^{n} {}^n c_k{}^{n-k} c_{j-k} (-1)^{j+k} \mu_j$$

$$= {}^n c_k \sum_{j=k}^{n} {}^{n-k} c_{j-k} (-1)^{j+k} \mu_j = {}^n c_k \sum_{j=0}^{n-k} {}^{n-k} c_j (-1)^{j+2k} \mu_{j+k}$$

$$= {}^n c_k \sum_{j=0}^{n-k} {}^{n-k} c_j (-1)^{j} \mu_{j+k} = {}^n c_k (\Delta^{n-k} \mu)_k,$$

completing the proof.

Theorem 3.19 If $(H, \mu) = (h_{nk})$ is a Hausdorff method, then

$$\sum_{k=0}^{n} h_{nk} = \mu_0$$

for all n.

Proof: We have, for all n,

$$\sum_{k=0}^{n} h_{nk} = \sum_{k=0}^{n} \sum_{j=k}^{n} {}^n c_j^{\,j} c_k (-1)^{j+k} \mu_j$$

$$= \sum_{j=0}^{n} {}^n c_j (-1)^{j} \left[\sum_{k=0}^{j} {}^j c_k (-1)^{k} \right] \mu_j$$

$$= \mu_0 + \sum_{j=1}^{n} {}^n c_j (-1)^{j} \left[\sum_{k=0}^{j} {}^j c_k (-1)^{k} \right] \mu_j = \mu_0,$$

since $\sum_{k=0}^{j} {}^j c_k (-1)^{k} = 0$, completing the proof.

Theorem 3.20   The following statements are equivalent:

(i) There exist totally monotone sequences $\{\alpha_n\}$, $\{\beta_n\}$, such that $\mu_n = \alpha_n - \beta_n$, for all n.

(ii) There exists an $M > 0$, such that

$$\sup_{n \geq 0} \left\{ \sum_{k=0}^{n} {}^n c_k |(\Delta^{n-k} \mu)_k| \right\} \leq M.$$

Proof: We first prove that (i) implies (ii). Let $\mu_n = \alpha_n - \beta_n$ for all n, where $\{\alpha_n\}$, $\{\beta_n\}$ are totally monotone sequences. Consequently,

$$\sum_{k=0}^{n} {}^n c_k |(\Delta^{n-k} \mu)_k| = \sum_{k=0}^{n} {}^n c_k \left| \sum_{j=0}^{n-k} {}^{n-k} c_j (-1)^j \mu_{k+j} \right|$$

$$= \sum_{k=0}^{n} {}^n c_k \left| \sum_{j=0}^{n-k} {}^{n-k} c_j (-1)^j (\alpha_{k+j} - \beta_{k+j}) \right|$$

$$= \sum_{k=0}^{n} {}^n c_k \left| \sum_{j=0}^{n-k} {}^{n-k} c_j (-1)^j \alpha_{k+j} - \sum_{j=0}^{n-k} {}^{n-k} c_j (-1)^j \beta_{k+j} \right|$$

$$= \sum_{k=0}^{n} {}^n c_k |(\Delta^{n-k} \alpha)_k - (\Delta^{n-k} \beta)_k| \leq \sum_{k=0}^{n} {}^n c_k [|(\Delta^{n-k} \alpha)_k| + |(\Delta^{n-k} \beta)_k|]$$

$$= \sum_{k=0}^{n} {}^n c_k (\Delta^{n-k} \alpha)_k + \sum_{k=0}^{n} {}^n c_k (\Delta^{n-k} \beta)_k = \alpha_0 + \beta_0.$$

Thus, (ii) holds with $M = \alpha_0 + \beta_0$. We now prove that (ii) implies (i). Let (ii) hold. Now

$$|(\Delta^m \mu)_p| = |\{(\Delta^m \mu)_p - (\Delta^m \mu)_{p+1}\} + (\Delta^m \mu)_{p+1}|$$

$$= |(\Delta^{m+1} \mu)_p + (\Delta^m \mu)_{p+1}|$$

$$\leq |(\Delta^{m+1} \mu)_p| + |(\Delta^m \mu)_{p+1}|.$$

Continuing this argument, we have

$$|(\Delta^m \mu)_p| \leq \sum_{j=0}^{k} {}^k c_j |(\Delta^{m+k-j} \mu)_{p+j}| = D_{m,p}^{(k)}.$$

Let,

$$M_n = \sum_{k=0}^{n} {}^n c_k |(\Delta^{n-k} \mu)_k|, \text{ for all } n.$$

We note that $D_{0,0}^{(k)} = M_k$ and $D_{m,p}^{(k)} \leq D_{m,p}^{(k+1)}$ for all m, p, and so $\{M_k\}$ is a nondecreasing sequence bounded above by M and thus is convergent. Further, for all m and p, we can write

$$D_{m,p}^{(k)} = \sum_{j=0}^{k} {}^{k}c_j |(\Delta^{m+k-j}\mu)_{p+j}| \leq \sum_{j=0}^{k} {}^{(m+p+k)}c_{(p+j)} |(\Delta^{m+k-j}\mu)_{p+j}|$$

$$= \sum_{j=0}^{p+k} {}^{(m+p+k)}c_j |(\Delta^{m+p+k-j}\mu)_j| = M_{m+p+k} \leq M.$$

So, $\{D_{m,p}^{(k)}\}$ is a nondecreasing sequence in k, bounded above by M and thus it is convergent. Let

$$\sigma_{m,p} = \lim_{k \to \infty} D_{m,p}^{(k)}.$$

Then $\sigma_{m,p} \geq 0$. Note that

$$D_{m,p}^{(k)} \leq D_{m+1,p}^{(k)} + D_{m,p+1}^{(k)} \leq D_{m,p}^{(k+1)}.$$

Hence, taking the limit as $k \to \infty$, we have

$$\sigma_{m,p} \leq \sigma_{m+1,p} + \sigma_{m,p+1} \leq \sigma_{m,p}.$$

Consequently,

$$\sigma_{m+1,p} + \sigma_{m,p+1} = \sigma_{m,p};$$
i.e., $$\sigma_{m+1,p} = \sigma_{m,p} - \sigma_{m,p+1}.$$

Let $\sigma_{0,p} = \sigma_p$, for all p. Since

$$(\Delta^m \sigma)_p = \sigma_{m,p} \geq 0,$$

$\{\sigma_p\}$ is totally monotone. Define the sequences $\alpha = \{\alpha_p\}$, $\beta = \{\beta_p\}$ by

$$\alpha_p = \frac{1}{2}(\sigma_p + \mu_p),$$

$$\beta_p = \frac{1}{2}(\sigma_p - \mu_p).$$

Now

$$(\Delta^m \sigma)_p = \sigma_{m,p} \geq D_{m,p}^{(k)} \geq |(\Delta^m \mu)_p|,$$

so α and β are totally monotone, and, further

$$\mu_p = \alpha_p - \beta_p.$$

Theorem 3.21 Let $\mu = \{\mu_k\}$ be a sequence of real numbers. If

$$\sup_{n \geq 0} \left\{ \sum_{k=0}^{n} {}^{n}c_k |(\Delta^{n-k}\mu)_k| \right\} \leq M, M > 0,$$

then

$$\lim_{n\to\infty} {}^n c_k (\Delta^{n-k}\mu)_k = 0, k = 1, 2, \ldots.$$

Proof: Using Theorem 3.20, μ is the difference of two totally monotone sequences. Thus, it suffices to assume that μ is totally monotone. Now

$$(\Delta^{n-k}\mu)_k = (\Delta^{n-k+1}\mu)_k + (\Delta^{n-k}\mu)_{k+1}.$$

So,

$${}^n c_k (\Delta^{n-k}\mu)_k = {}^n c_k (\Delta^{n-k+1}\mu)_k + {}^n c_k (\Delta^{n-k}\mu)_{k+1},$$

and

$$(n+1)^n c_k (\Delta^{n-k}\mu)_k = (n-k+1)^{(n+1)} c_k (\Delta^{n-k+1}\mu)_k + (k+1)^{(n+1)} c_{(k+1)} (\Delta^{n-k}\mu)_{k+1};$$

that is,

$$
\begin{aligned}
(n+1)[{}^n c_k (\Delta^{n-k}\mu)_k &- {}^{(n+1)} c_k (\Delta^{n-k+1}\mu)_k] \\
&= (k+1)^{(n+1)} c_{(k+1)} (\Delta^{n-k}\mu)_{k+1} \\
&+ [(n-k+1)^{(n+1)} c_k (\Delta^{n-k+1}\mu)_k - (n+1)^{(n+1)} c_k (\Delta^{n-k+1}\mu)_k] \\
&= (k+1)^{(n+1)} c_{(k+1)} (\Delta^{n-k}\mu)_{k+1} - k^{(n+1)} c_k (\Delta^{n-k+1}\mu)_k.
\end{aligned}
$$

This implies,

$$
\begin{aligned}
(n+1) \sum_{j=0}^{k} [{}^n c_j (\Delta^{n-j}\mu)_j &- {}^{(n+1)} c_j (\Delta^{n-j+1}\mu)_j] \\
&= \sum_{j=0}^{k} [(j+1)^{(n+1)} c_{(j+1)} (\Delta^{n-j}\mu)_{j+1} - j^{(n+1)} c_j (\Delta^{n-j+1}\mu)_j] \\
&= (n+1)^n c_k (\Delta^{n-k}\mu)_{k+1} \geq 0.
\end{aligned}
$$

So,

$$\left\{ \sum_{j=0}^{k} {}^n c_j (\Delta^{n-j}\mu)_j \right\}$$

is a decreasing sequence of nonnegative numbers in n. Thus, this sequence converges to a limit. Hence,

$$\lim_{n\to\infty} {}^n c_k (\Delta^{n-k}\mu)_k = \lim_{n\to\infty} \left[\sum_{j=0}^{k} {}^n c_j (\Delta^{n-j}\mu)_j - \sum_{j=0}^{k-1} {}^n c_j (\Delta^{n-j}\mu)_j \right]$$

exists finitely (say) $= t_k$. Let

$$p_n = \sum_{j=0}^{k} {}^{n}c_j(\Delta^{n-j}\mu)_j - \sum_{j=0}^{k} {}^{(n+1)}c_j(\Delta^{n-j+1}\mu)_j,$$

for all n. We have proved that

$$(n+1)p_n = (n+1)^n c_k(\Delta^{n-k}\mu)_{k+1}$$

$$= (k+1)^{(n+1)}c_{(k+1)}(\Delta^{n-k}\mu)_{k+1}.$$

Thus,

$$p_n = \frac{k+1}{n+1}{}^{(n+1)}c_{(k+1)}(\Delta^{n-k}\mu)_{k+1}$$

$$\sim \frac{k+1}{n+1}t_{k+1}, n \to \infty.$$

Note that $\Sigma_n p_n$ converges, and so it follows that

$$t_{k+1} = 0, \text{ for all } k; \text{ i.e., } \lim_{n\to\infty} {}^{n}c_k(\Delta^{n-k}\mu)_k = 0, k = 1, 2, \ldots .$$

Theorem 3.22 A Hausdorff method (H, μ) is regular if and only if

(i) $\lim_{n\to\infty} (\Delta^n\mu)_0 = 0$;

(ii) $\mu_0 = 1$;

and

(iii) μ is the difference of two totally monotone sequences.

Proof: In view of Theorem 1.1, $(H, \mu) = (h_{nk})$ is regular if and only if

$$\lim_{n\to\infty} h_{nk} = 0; \tag{3.8}$$

$$\lim_{n\to\infty} \sum_k h_{nk} = 1; \tag{3.9}$$

and

$$\sup_{n\geq 0} \sum_k |h_{nk}| < \infty. \tag{3.10}$$

Let (H, μ) be regular. In view of (3.8),

$$\lim_{n\to\infty} (\Delta^n\mu)_0 = 0.$$

By (3.9) and Theorem 3.19, we can assert that $\mu_0 = 1$. Using (3.10) and Theorem 3.20, we have that μ is the difference of two totally monotone sequences. Hence (i), (ii), and (iii) hold.

Conversely, let (i), (ii), and (iii) hold. Then, (3.10) holds by (iii) and Theorem 3.20, and (3.9) holds by (ii) and Theorem 3.19. In view of (i), (iii), Theorem 3.20, and Theorem 3.21, it follows that (3.8) holds. Consequently, (H, μ) is regular, completing the proof of the theorem.

3.6 Excercise

Exercise 3.1 Prove that the converse of Theorem 3.6 is not true.

Exercise 3.2 Prove that the Borel matrix method of summability is regular.

Exercise 3.3 Prove that the Borel exponential method is included in the Borel matrix method.

Exercise 3.4 Prove that the (H, m) method is regular.
Hint. See [1].

Exercise 3.5 Prove that if $m > n$, m, n being positive integers, then
$$(H, n) \subsetneqq (H, m).$$
Hint. See [1].

Exercise 3.6 Prove that if $\alpha > 0$, then $(C, \alpha) = (c_{kj}^{(\alpha)})$ is regular.
Hint. See [1].

Exercise 3.7 Prove that if $\alpha > -1$ and $h > 0$, then
$$(C, \alpha) \subsetneqq (C, \alpha + h).$$
Hint. See [1].

Exercise 3.8 Give an example of an (M, λ_n) method, which is (i) regular and (ii) not regular.

Exercise 3.9 If $\sum_n a_n$ is (C, k) summable, $k > -1$, prove that $a_n = o(n^k)$, $n \to \infty$. (Limitation theorem for Cesàro method (C, k).)
Hint. See [1].

Exercise 3.10 If $\sum_n a_n = s$ and $na_n = O(1)$, $n \to \infty$, prove that $\sum_n a_n = s(C, -1 + \delta)$, for $\delta > 0$.
Hint. See [1, Theorem 45].

Exercise 3.11 If $\sum_n a_n$ is (C, k) summable to s, $k > -1$, prove that $\sum_n a_n$ is Abel summable to s but not conversely.
Hint. See [1].

Exercise 3.12 Show that neither $(C, 1) \subseteq (E, 1)$ nor $(E, 1) \subseteq (C, 1)$ is valid.

Exercise 3.13 Is the Nörlund method translative? Justify your answer.

Exercise 3.14 Is the (C, k) method, $k > -1$, a Hausdorff method? Why?

Exercise 3.15 Prove that $\{1/n + 1\}$ is totally monotone.

Exercise 3.16 Prove that $\{r^n\}$ is totally monotone for $0 < r \leq 1$.

References

1 Hardy, G.H.: Divergent Series. Oxford University Press, Oxford (1949).
2 Natarajan, P.N.: On the (M, λ_n) of summability. Analysis (München) **33**, 51–56 (2013).
3 Powell, R.E. and Shah, S.M.: Summability Theory and Applications. Prentice-Hall of India, Delhi (1988).

4

Tauberian Theorems

4.1 Brief Introduction

In 1897, A. Tauber proved the following theorem:
If a real series $\Sigma_n a_n$ is Abel summable to s and if $a_n = o(1/n)$, $n \to \infty$, then the series $\Sigma_n a_n$ converges to s. Theorems of this type are named after Tauber and are called Tauberian theorems. In such theorems, we conclude that $\Sigma_n a_n$ converges to s given that $\Sigma_n a_n$ is summable to s by some regular summability method A, with an additional condition on the terms a_n of the given series. Throughout this chapter, we suppose that the a_n's are real. We note that most of the theorems are true when the a_n's are complex. For instance, we prove the following theorem when the a_n's are real: If $\Sigma_n a_n$ is $(C, 1)$ summable to s and if $a_n = O(1/n)$, $n \to \infty$, then $\Sigma_n a_n$ converges to s. This theorem continues to hold when a_n's are complex. Let $a_n = \alpha_n + i\beta_n$, $n = 0, 1, 2, \dots$. If $a_n = O(1/n)$, $n \to \infty$, then

$$\alpha_n = O\left(\frac{1}{n}\right), \beta_n = O\left(\frac{1}{n}\right), n \to \infty.$$

So, $\Sigma_n \alpha_n$, $\Sigma_n \beta_n$ both converge and hence, $\Sigma_n a_n = \Sigma_n(\alpha_n + i\beta_n)$ converges. Since the method A is regular, it follows that $\Sigma_n a_n$ converges to s. In this chapter, we study some elementary Tauberian theorems only, leaving the study of deeper Tauberian theorems to the reader.

4.2 Tauberian Theorems

We state the following result (for details of proof see [1, 2]).

Theorem 4.1 Let $\alpha > -1$. If $\Sigma_n a_n$ is (C, α)-summable to s, then $\Sigma_n a_n$ is Abel-summable to s but not conversely,

i.e., $(C, \alpha) \subsetneq (A)$.

An Introductory Course in Summability Theory, First Edition. Ants Aasma, Hemen Dutta, and P.N. Natarajan.
© 2017 John Wiley & Sons, Inc. Published 2017 by John Wiley & Sons, Inc.

It follows that any Tauberian theorem for the Abel method continues to be a Tauberian theorem for the (C, α) method, $\alpha > -1$. We have a similar result for the Natarajan method (M, λ_n), since, we have seen earlier that

$$(M, \lambda_n) \subsetneq (A).$$

We need the following lemma to prove our next important theorem.

Lemma 4.1   If $\Sigma_n a_n x^n$ converges for $|x| < 1$ and $\Sigma_n a_n = \infty$, then

$$\lim_{x \to 1-} \sum_n a_n x^n = \infty.$$

Proof: Let $f(x) = \Sigma_n a_n x^n$ for $|x| < 1$. Then, for $|x| < 1$,

$$f(x) = f(x)(1 - x)\left(\sum_n x^n \right)$$

$$= (1 - x)\left(\sum_n x^n \right)\left(\sum_n a_n x^n \right)$$

$$= (1 - x)\left(\sum_n s_n x^n \right),$$

where $s_n = \sum_{k=0}^{n} a_k$ for all n. Since $\lim_{n \to \infty} s_n = \infty$, given an $H > 0$, there exists a positive integer n_1, such that

$$S_n > H, n > n_1.$$

We can now choose x_1, such that if $0 < x_1 < x < 1$, then

$$x^{n_1 + 1} > \frac{1}{2}.$$

Also, we can choose x_2, such that if $0 < x_2 < x < 1$, then

$$1 - x < \frac{H}{4\left(\sum_{n=0}^{n_1} |s_n| \right)}.$$

Consequently, for $0 < \max(x_1, x_2) < x < 1$, we have

$$f(x) = (1 - x)\left(\sum_{n=0}^{n_1} s_n x^n + \sum_{n=n_1+1}^{\infty} s_n x^n \right).$$

Thus,

$$(1 - x)\left(\sum_{n=n_1+1}^{\infty} s_n x^n \right) > H(1 - x)\left(\sum_{n=n_1+1}^{\infty} x^n \right).$$

$$= H(1-x)\frac{x^{n_1+1}}{(1-x)} = Hx^{n_1+1} > \frac{H}{2};$$

$$(1-x)\left|\sum_{n=0}^{n_1} s_n x^n\right| < (1-x)\sum_{n=0}^{n_1} |s_n| < \frac{H}{4}.$$

Consequently, for the above choice of x with $0 < x < 1$, we have

$$f(x) > (1-x)\sum_{n=n_1+1}^{\infty} s_n x^n - (1-x)\left|\sum_{n=0}^{n_1} s_n x^n\right| > \frac{H}{2} - \frac{H}{4} = \frac{H}{4}.$$

Since H is arbitrary, it follows that

$$\lim_{x\to 1-}\sum_n a_n x^n = \lim_{x\to 1-} f(x) = \infty,$$

completing the proof of the lemma. □

We now have the following Tauberian theorem.

Theorem 4.2 If $\Sigma_n a_n$ is Abel-summable to s, and $a_n \geq 0$ for sufficiently large n, then $\Sigma_n a_n$ converges to s.

Proof: By hypothesis, there exists a positive integer n_0 such that

$$a_n \geq 0, n > n_0.$$

So $\{s_n\}$ is a nondecreasing sequence, where $s_n = \sum_{k=0}^{n} a_k$ for all n. If $s_n \neq O(1)$, $n \to \infty$, then $\lim_{n\to\infty} s_n = \infty$. In view of Lemma 4.1, $\Sigma_n a_n$ is not Abel-summable, which is a contradiction. Thus, $s_n = O(1)$, $n \to \infty$. Hence, $\{s_n\}$, being a nondecreasing sequence, which is bounded, converges to L (say). Since the Abel method is regular, by Theorem 2.9, $\{s_n\}$ is Abel-summable to L, and so $s = L$.

In view of Theorems 4.1 and 4.2, we have the following result.

Corollary 4.1 Let $\alpha > -1$. If $\Sigma_n a_n$ is (C, α)-summable to s and $a_n \geq 0$ for sufficiently large n, then $\Sigma_n a_n$ converges to s.

Example 4.1 We show that if $\Sigma_n a_n$ is $(C, 1)$ summable to s and $a_n \geq 0$ for sufficiently large n, then $\Sigma_n a_n$ converges to s. Indeed, if $\Sigma_n a_n$ is $(C, 1)$ summable to s, then it is Abel summable to s. By Theorem 4.2, it follows that $\Sigma_n a_n$ converges to s.

The following is an interesting result in itself.

Theorem 4.3 Let $T_n = \sum_{j=1}^{n} j a_j$. If $\Sigma_n a_n$ is $(C, 1)$-summable to s, then $\Sigma_n a_n$ converges to s if and only if

$$T_n = o(n), n \to \infty.$$

Proof: We recall that

$$S_n^{(0)} = S_n = \sum_{j=0}^{n} a_j,$$

$$S_n^{(1)} = \sum_{j=0}^{n} S_j, \text{ for all } n. \text{ Now}$$

$$(n+1)S_n - S_n^{(1)} = (n+1)\sum_{j=0}^{n} a_j - \sum_{j=0}^{n} S_j = \sum_{j=0}^{n} j a_j = T_n.$$

Therefore,

$$S_n - \frac{S_n^{(1)}}{n+1} = \frac{T_n}{n+1}. \tag{4.1}$$

If $\Sigma_n a_n$ converges to s, $\lim_{n\to\infty} S_n = s$. Since $(C, 1)$-summability is regular, then

$$\lim_{n\to\infty} \frac{S_n^{(1)}}{n+1} = s.$$

Consequently, by (4.1), we obtain

$$T_n = o(n), \ n \to \infty.$$

Conversely, if $T_n = o(n)$ as $n \to \infty$, then

$$\lim_{n\to\infty} \left[S_n - \frac{S_n^{(1)}}{n+1} \right] = 0$$

by (4.1). By hypothesis, $\Sigma_n a_n$ is $(C, 1)$-summable to s and so,

$$\lim_{n\to\infty} \frac{S_n^{(1)}}{n+1} = s.$$

Hence, $\lim_{n\to\infty} S_n = s$, that is, $\Sigma_n a_n$ converges to s.

Corollary 4.2 If $\Sigma_n a_n$ is $(C, 1)$-summable to s and $a_n = o(1/n)$, $n \to \infty$, then $\Sigma_n a_n$ converges to s.

Proof: Since $n a_n = o(1)$, $T_n = o(n)$ by the regularity of the $(C, 1)$ method. In view of Theorem 4.3, $\Sigma_n a_n$ converges to s.

Theorem 4.4 If $\Sigma_n a_n$ is Abel-summable to s, then $\Sigma_n a_n$ converges to s if and only if $T_n = o(n)$ as $n \to \infty$, where $T_n = \sum_{j=0}^{n} j a_j$ for all n.

Proof: We first suppose that

$$na_n = o(1), \ n \to \infty.$$

If $0 < x < 1$, then

$$f(x) - S_m = \sum_n a_n x^n - \sum_{n=0}^m a_n$$

$$= \sum_{n=0}^m a_n x^n - \sum_{n=0}^m a_n + \sum_{n=m+1}^\infty a_n x^n$$

$$= \sum_{n=m+1}^\infty a_n x^n - \sum_{n=0}^m a_n (1 - x^n).$$

Now

$$1 - x^n = (1 - x) \left(\sum_{j=0}^{n-1} x^j \right) < n(1 - x)$$

since $x < 1$. Let

$$\epsilon_m = \max_{k \geq m} |k a_k|,$$

for all m. Note that $\{\epsilon_n\}$ decreases to 0 as $n \to \infty$. Now

$$|f(x) - S_m| \leq \sum_{n=1}^m (1 - x) n |a_n| + \left| \sum_{n=m+1}^\infty \frac{n a_n x^n}{n} \right|$$

$$< (1 - x) \sum_{n=1}^m n |a_n| + \epsilon_{m+1} \sum_{n=m+1}^\infty \frac{x^n}{n}$$

$$< (1 - x) \sum_{n=1}^m n |a_n| + \frac{\epsilon_{m+1}}{m + 1} \sum_{n=m+1}^\infty x^n$$

$$\leq (1 - x) \sum_{n=1}^m n |a_n| + \frac{\epsilon_{m+1}}{m + 1} \sum_n x^n$$

$$= (1 - x) \sum_{n=1}^m n |a_n| + \frac{\epsilon_{m+1}}{m + 1} \frac{1}{1 - x},$$

since $0 < x < 1$. By assumption, $na_n = o(1)$ as $n \to \infty$ and so

$$\sum_m na_n = o(m), \ m \to \infty.$$

We recall that $\{\epsilon_m\}$ decreases to 0 as $m \to \infty$. Therefore,

$$\left| f\left(1 - \frac{1}{m}\right) - S_m \right| \leq \frac{1}{m} \sum_{n=0}^m n |a_n| + \frac{m}{m + 1} \epsilon_{m+1} \to 0, \ m \to \infty;$$

that is,

$$\lim_{m \to \infty} \left[f\left(1 - \frac{1}{m}\right) - S_m \right] = 0.$$

Since $\Sigma_n a_n$ is Abel-summable to s,

$$\lim_{m \to \infty} f\left(1 - \frac{1}{m}\right) = s.$$

Thus, $\lim_{m \to \infty} S_m = s$, that is, $\sum_n a_n$ converges to s.

Let now $T_n = o(n)$ as $n \to \infty$. Note that,

$$T_0 = 0, \quad \frac{T_n - T_{n-1}}{n} = a_n, \quad n = 1, 2, \cdots.$$

Hence, for $0 < x < 1$, we can write

$$f(x) = a_0 + \sum_{n=1}^{\infty} \left(\frac{T_n - T_{n-1}}{n} \right) x^n$$

$$= a_0 + \sum_{n=1}^{\infty} \frac{T_n}{n} x^n - \sum_{n=1}^{\infty} \frac{T_{n-1}}{n} x^n$$

$$= a_0 + \sum_{n=1}^{\infty} \frac{T_n}{n} x^n - \sum_{n} \frac{T_n}{n+1} x^{n+1}$$

$$= a_0 + \sum_{n=1}^{\infty} \frac{T_n}{n} x^n - \sum_{n=1}^{\infty} \frac{T_n}{n+1} x^{n+1}$$

$$= a_0 + \sum_{n=1}^{\infty} \frac{T_n}{n} x^n - \sum_{n=1}^{\infty} \left\{ \frac{1}{n} - \frac{1}{n(n+1)} \right\} T_n x^{n+1}$$

$$= a_0 + (1-x) \sum_{n=1}^{\infty} \frac{T_n}{n} x^n + \sum_{n=1}^{\infty} \frac{T_n}{n(n+1)} x^{n+1}.$$

Let $\delta_m = \max_{k \geq m} |T_n|/n$, for all m. Note that $\{\delta_m\}$ decreases to 0 as $m \to \infty$. Also,

$$\overline{\lim_{x \to 1-}} \left| (1-x) \sum_{n=1}^{\infty} \frac{T_n}{n} x^n \right|$$

$$\leq \overline{\lim_{x \to 1-}} (1-x) \left(\sum_{n=1}^{m} \frac{|T_n|}{n} x^n \right) + \overline{\lim_{x \to 1-}} (1-x) \frac{\delta_{m+1}}{(1-x)}$$

$$\leq \overline{\lim_{x \to 1-}} (1-x) \left(\sum_{n=1}^{m} \frac{|T_n|}{n} x^n \right) + \delta_{m+1},$$

from which it follows that

$$\lim_{x \to 1-} (1-x) \sum_{n=1}^{\infty} \frac{T_n}{n} x^n = 0.$$

Consequently,

$$\lim_{x \to 1-} f(x) = a_0 + \lim_{x \to 1-} \sum_{n=1}^{\infty} \frac{T_n}{n(n+1)} x^{n+1}$$

$$= a_0 + \lim_{x \to 1-} \sum_{n=1}^{\infty} \frac{T_n}{n(n+1)} x^n;$$

that is,

$$\lim_{x \to 1-} \left(\sum_n a_n x^n \right) = a_0 + \lim_{x \to 1-} \left(\sum_{n=1}^{\infty} u_n x^n \right),$$

where

$$u_n = \frac{T_n}{n(n+1)}, \ n = 1, 2, \dots .$$

This implies that

$$\lim_{x \to 1-} (1-x) \left(\sum_n S_n x^n \right) = a_0 + \lim_{x \to 1-} (1-x) \left(\sum_n U_n x^n \right),$$

where

$$S_n = \sum_{k=0}^{n} a_k, \ U_n = \sum_{k=0}^{n} u_k.$$

Hence,

$$s = a_0 + \lim_{x \to 1-} (1-x) \left(\sum_n U_n x^n \right),$$

since $\sum_n a_n$ is Abel-summable to s, that is,

$$\sum_n u_n = \sum_{n=1}^{\infty} \frac{T_n}{n(n+1)}$$

is Abel-summable to $s - a_0$. Now

$$n u_n = n \frac{T_n}{n(n+1)} = \frac{T_n}{n+1} = o(1), n \to \infty,$$

since $T_n = o(n)$, as $n \to \infty$. By the first part of the proof, it follows that

$$\sum_{n=1}^{\infty} u_n = \sum_{n=1}^{\infty} \frac{T_n}{n(n+1)}$$

converges to $s - a_0$.

However,

$$\sum_{n=1}^{\infty} \frac{T_n}{n(n+1)} = \lim_{n \to \infty} \sum_{k=1}^{n} \frac{T_k}{k(k+1)}$$

$$= \lim_{n \to \infty} \sum_{k=1}^{n} T_k \left(\frac{1}{k} - \frac{1}{k+1} \right)$$

$$= \lim_{n \to \infty} \left(\sum_{k=1}^{n} \frac{T_k}{k} - \sum_{k=1}^{n} \frac{T_k}{k+1} \right)$$

$$= \lim_{n \to \infty} \left(\sum_{k=1}^{n} \frac{T_k}{k} - \sum_{k=2}^{n+1} \frac{T_{k-1}}{k} \right)$$

$$= \lim_{n \to \infty} \left(\sum_{k=1}^{n} \frac{T_k}{k} - \sum_{k=1}^{n+1} \frac{T_{k-1}}{k} \right)$$

$$= \lim_{n \to \infty} \left(\sum_{k=1}^{n} \frac{T_k - T_{k-1}}{k} - \frac{T_n}{n+1} \right)$$

$$= \lim_{n \to \infty} \left(\sum_{k=1}^{n} a_k - \frac{T_n}{n+1} \right)$$

$$= \sum_{k=1}^{n} a_k,$$

since $T_n = o(n)$ as $n \to \infty$, that is,

$$\sum_{k=1}^{n} a_k \text{ converges to } s - a_0;$$

that is,

$$\sum_{k} a_k \text{ converges to } s.$$

Corollary 4.3   In view of Theorem 4.1 we have the following: Let $\alpha > -1$. If $\sum_n a_n$ is (C, α) summable to s and $T_n = o(n)$, then $\sum_n a_n$ converges to s.

Corollary 4.4   If $\sum_n a_n$ is Abel-summable to s and $a_n = o(1/n)$, $n \to \infty$, then $\sum_n a_n$ converges to s.

In view of Theorem 4.1, Corollary 4.2 follows immediately from Corollary 4.4.

Example 4.2   We   show   that   if $\sum_n a_n$ is (M, λ_n) summable to s and $a_n = o(1/n)$ as $n \to \infty$, then $\sum_n a_n$ converges to s. Indeed, if $\sum_n a_n$ is (M, λ_n) summable to s, then it is Abel summable to s. By Corollary 4.4, it follows that $\sum_n a_n$ converges to s.

Example 4.3 (Hardy–Littlewood theorem) If $\sum_n a_n$ is Abel summable to s and $na_n \geq -c$, for some $c > 0$ and for all n, then $\sum_n a_n$ converges to s. Indeed, since $na_n \geq -c$, it follows that

$$\frac{1}{n} \sum_{k=0}^{n} ka_k \geq -c.$$

Using Exercise 4.5, $\sum_n a_n$ is $(C, 1)$ summable to s. Using Exercise 4.3, it follows that $\sum_n a_n$ converges to s.

4.3 Excercise

Exercise 4.1   Prove that

$$\sum_n a_n \text{ is } (C, 1)$$

summable if and only if

$$\sum_{n=1}^{\infty} T_n/n(n+1)$$

converges, where, as usual,

$$T_n = \sum_{j=1}^{n} ja_j, n = 1, 2, \ldots.$$

Exercise 4.2   Prove that if $\sum_n a_n$ is $(C, 1)$-summable to s and $a_n = O(1/n)$, $n \to \infty$, then $\sum_n a_n$ converges to s.

Hint. See [2, pp. 77–78.]

Exercise 4.3   If $\sum_n a_n$ is $(C, 1)$ summable to s and $na_n \geq -c$, for some $c > 0$ and for all n, prove that $\sum_n a_n$ converges to s.

Hint. See [2, pp. 79–80.]

Exercise 4.4   If $\sum_n a_n$ is Abel summable to s and

$$s_n = \sum_{k=0}^{n} a_k \geq -c,$$

for some $c > 0$ and for all n, prove that $\sum_n a_n$ is $(C, 1)$ summable to s.

Hint. See [2, pp. 82–83.]

Exercise 4.5   If $\Sigma_n a_n$ is Abel summable to s and

$$\frac{1}{n} \sum_{k=0}^{n} k a_k \geq -c,$$

for some $c > 0$ and for all n, prove that $\Sigma_n a_n$ is $(C, 1)$ summable to s.

Hint. See [2, pp. 83–84.]

References

1 Hardy, G.H.: Divergent Series. Oxford University Press, Oxford (1949).
2 Powell, R.E. and Shah, S.M.: Summability Theory and Applications. Prentice-Hall of India, Delhi (1988).

5

Matrix Transformations of Summability and Absolute Summability Domains: Inverse-Transformation Method

5.1 Introduction

Let ω be the set of all sequences over the real or complex numbers, and X, Y be subsets of ω. Let $M = (m_{nk})$ be a matrix with real or complex entries. Throughout this chapter, we assume that all indices and summation indices run from 0 to ∞, unless otherwise specified. Also we note at the beginning that all the notions and notations not defined in this chapter can be found in Chapter 1. However, instead of $x := \{x_k\}$, we denote a sequence in this chapter by $x := (x_k)$. If $Mx := (M_n x) \in Y$ for every $x = (x_k) \in Y$, where $Mx := M(x) = \{(Mx)_n\}$ and

$$M_n x := (Mx)_n = \sum_k m_{nk} x_k,$$

then, we write $M \in (X, Y)$. In that case we say that M transforms X into Y. Let $A = (a_{nk})$ and $B = (b_{nk})$ be matrices with real or complex entries. Then, we denote the summability and the absolute summability domains of A, correspondingly, by c_A (or cs_A) and bv_A (or l_A), that is,

$$cs_A := \{x = (x_k) \in \omega \ : \ (A_n x) \in cs\},$$
$$c_A := \{x = (x_k) \in \omega \ : \ (A_n x) \in c\},$$
$$bv_A := \{x = (x_k) \in \omega, \ : \ (A_n x) \in bv\},$$
$$l_A := \{x = (x_k) \in \omega \ : \ (A_n x) \in l\}.$$

Thus, a matrix A determines a summability method on c_A, cs_A, bv_A, or l_A. Therefore, instead of "matrix," we may sometimes write "method." Besides, under the terms "matrix" and "method," we mean, further, a method or a matrix with real or complex entries, if not specified otherwise. In this chapter, we describe necessary and sufficient conditions for $M \in (c_A, c_B)$, $M \in (cs_A, cs_B)$, $M \in (bv_A, c_B)$, and $M \in (bv_A, bv_B)$.

This problem was studied for the first time by Alpár in 1978 (see [13]). He found necessary and sufficient conditions for M to be a transform from cs into c_{C^α} for $\alpha \geq 0$. Later (see [13, 14]), he generalized this result by establishing

An Introductory Course in Summability Theory, First Edition. Ants Aasma, Hemen Dutta, and P.N. Natarajan.
© 2017 John Wiley & Sons, Inc. Published 2017 by John Wiley & Sons, Inc.

necessary and sufficient conditions for M to be a transform from c_{C^α} into c_{C^β} for $\alpha \geq 0, \beta \geq 0$. In [13–15] Alpár applied the above-mentioned results for studying the conformal mapping in the theory of complex functions. In 1986 (see [40]), Thorpe generalized Alpár's results taking, instead of C^β, an arbitrary normal method B. Further, Aasma generalized Alpár's and Thorpe's results (see [1–5, 7–11]). In 1987 (see [1]), he found necessary and sufficient conditions for M to be a transform from c_A into c_B, and, in 1994 (see [5]) from bv_A into c_B, and from bv_A into bv_B for a reversible A and a triangular B. Sufficient conditions for the transforms from c_A into c_B, from bv_A into c_B and from bv_A into bv_B for a reversible A and an arbitrary B are presented in [2]. Necessary and sufficient conditions for M to be a transform from cs_A into cs_B for a normal method A and a triangular B are found in [11].

Necessary and sufficient conditions for M to be a transform from c_A into c_B, for a regular perfect A and a triangular B, were found by Aasma in 1994 (see [4]). Some classes of matrices M transforming c_A into c_B were found in [7–9].

We also consider the problem of M-consistency of matrix methods A and B on c_A, introduced in [4] and [11].

Definition 5.1 Matrices A and B is said to be M^{seq}-consistent on c_A if the transformation Mx exists and

$$\lim_n B_n(Mx) = \lim_n A_n x$$

for each $x \in c_A$, and M^{ser}-consistent on cs_A if the transformation Mx exists and

$$\lim S[B(Mx)] = \lim S(Ax)$$

for each $x \in cs_A$, where

$$Sx := (X_n), X_n := \sum_{k=0}^{n} x_k, \lim Sx := \lim_n X_n.$$

If

$$m_{nk} = \delta_{nk}\varepsilon_k, \tag{5.1}$$

where $\delta_{nn} = 1$ and $\delta_{nk} = 0$ for $n \neq k$, and (ε_k) is a sequence of numbers, then

$$B_s Mx = B_s(x_n\varepsilon_n)$$

for each $x \in c_A$ (bv_A). Hence, in this case, the above-mentioned problem reduces to the problem of finding necessary and sufficient conditions for numbers ε_k to be the summability (or the absolute summability) factors for matrix methods, defined by A and B. If B is the identity method E, that is, $b_{nk} = 1$ for all k and n, then we get necessary and sufficient conditions for numbers ε_k to be the convergence factors for the summability method, defined by A. The problem of convergence and summability factors have been widely investigated (see, e.g., [12, 16, 17, 20–23, 30, 32, 36, 37, 39]).

If, in addition, $\varepsilon_k = 1$ for every k, then $m_{nk} = \delta_{nk}$, that is, $Mx = x$ for each $x \in c_A$ (bv_A). So the inclusion problem of summability methods (see Definition 2.3) and M^{seq}-consistency (M^{ser}-consistency) of A and B on c_A (on cs_A) reduces to the problem of the usual consistency of A and B on c_A (see also Definition 2.2). For example, in this case necessary and sufficient conditions for $M \in (c_A, c_B)$ reduce to necessary and sufficient conditions for inclusion $A \subseteq B$, and necessary and sufficient conditions for M^{seq}-consistency of A and B on c_A reduce to necessary and sufficient conditions for the validity of the equation $\lim_n B_n x = \lim_n A_n x$ for every $x \in c_A$. The inclusion problem is also widely investigated (see, e.g., [26, 28, 29, 38]). A good overview on the problem of the summability factors and the problem of inclusion of matrix methods can be found in the monographs of Baron [16] and Leiger [35]. The inclusion of matrix methods is also described in the monographs of Boos [19] and Wilansky [44]. In the present work, we also describe conditions for inclusions $l_A \subset l_B$ and $l_A \subset cs_B$, and for M^{ser}-consistency of A and B on cs_A.

The results on the summability factors and inclusion theorems have been widely applied to summation of Fourier and orthogonal series. Good overviews of the problem of summation of Fourier and orthogonal series are given in the monographs of Butzer and Nessel (see [24]), Trebels (see [41]), and Zhuk (see [45]).

In this chapter, we consider only the cases where all matrices A, B, M, and sequences (ε_k) are defined over the real or complex numbers. Necessary and sufficient conditions for M to be a transform from c_A into c_B, for the case where the elements of M are continuous linear operators from a Banach space X into another Banach space Y, and the elements of A and B are continuous linear operators from X into X and from Y into Y, respectively, are given by Aasma (see [6]). The reader can find a similar kind of generalization for summability factors and inclusion of matrix methods, for example, from sources [32–34, 42, 43].

5.2 Some Notions and Auxiliary Results

In this section, we introduce some well-known notions and results (which a reader can find, e.g., from [16, 19, 27, 35, 44]), which will be required later.

Definition 5.2 A matrix A is called reversible if the infinite system of equations $z_n = A_n x$ has a unique solution for each sequence $(z_n) \in c$.

Definition 5.3 A sequence space X is called a BK-space if X is a Banach space, in which coordinatewise convergence holds.

Definition 5.4   A subset $M \subset X$ is called a fundamental set of X if its linear cover

$$\text{lin}\, M := \left\{ \sum_{k=0}^{n} \alpha_k x_k \; : \; \alpha_k \in \mathbf{C} \right\}$$

is dense in X.

Let

$$e := (1, 1, \cdots), e^k := (0, \cdots, 0, 1, 0, \cdots), \text{where } 1 \text{ is in the } k\text{th position,}$$

and $\Delta := \{e^0, e^1, \cdots\}$.

The space of continuous linear functionals on X (or the topological conjugate space of X) will be denoted by X'.

Lemma 5.1 Δ is a fundamental set of the set of null-sequences c_0 and $\Delta \cup e$ is a fundamental set of c. Every $x = (x_k) \in c$ can be uniquely represented in the form

$$x_k = \xi e + \sum_{l=0}^{k} (x_l - \xi) e^l; \; \xi := \lim_k x_k.$$

Lemma 5.2 c_0 and c are BK-spaces. The general form of a continuous linear functional $f \in c'$ can be presented by formula

$$f(x) = \alpha \xi + \sum_{k} \alpha_k x_k, \xi := \lim_k x_k, (\alpha_k) \in l, \alpha \in \mathbf{C}. \tag{5.2}$$

Let

$$c_A^0 := \{x = (x_k) \; : \; x \in c_A \text{ and } \lim_n A_n x = 0\}.$$

Lemma 5.3   If A is a reversible method, then c_A and c_A^0 are BK-spaces, where the norm $\|x\|$ is defined by the equality

$$\|x\| := \sup_n |A_n x|.$$

The general form of a continuous linear functional $f \in (c_A)'$ can be presented by formula

$$f(x) = \alpha \lim_n A_n x + \sum_n \alpha_n A_n x, (\alpha_n) \in l, \alpha \in \mathbf{C}. \tag{5.3}$$

Proof: We can consider A as a one-to-one mapping from c_A into c. Moreover, the mapping A, defined by the equality $y = Ax$ for every $x \in c_A$, and its inverse mapping A^{-1}, defined by the equality $x = A^{-1}y$ on c, are bounded (and therefore also continuous) by the equality

$$\|Ax\|_c = \|x\|_{c_A} \quad (x \in c_A).$$

Hence, c_A is isometrically isomorphic to the BK-space c. Similarly, we can show that c_A^0 is isometrically isomorphic to the BK-space c_0. Thus, c_A and c_A^0 are BK-spaces. Further, we show that $f \in (c_A)'$ if and only if f is representable in the form

$$f = g \cdot A; \; g \in c'. \tag{5.4}$$

Indeed, we can write $f = (f \cdot A^{-1}) \cdot A$. If we suppose that $f \in (c_A)'$, then $g := f \cdot A^{-1} \in c'$, since A^{-1} is a one-to-one continuous linear mapping from c into c_A. Conversely, if relation (5.4) holds, then clearly $f \in (c_A)'$. Hence the validity of (5.3) follows from (5.4) by (5.2).

Lemma 5.4 If A is a reversible method, then every element x_k of a sequence $x = (x_k) \in c_A$ can be represented in the form

$$x_k = \mu \eta_k + \sum_l \eta_{kl}(z_l - \mu); \; \sum_l |\eta_{kl}| < \infty, z_l := A_l x, \mu := \lim_l z_l, \tag{5.5}$$

where the sequences (η_{nr}) and (η_n) are the solutions of the system of equations $z_l = A_l x$ $(l = 0, 1, \cdots)$, correspondingly, to $z_l = \delta_{lr}$ and $z_l = \delta_{ll}$.

Proof: As c_A is a BK-space by Lemma 5.3, all elements x_k of $x = (x_k) \in c_A$ are continuous linear functionals on c_A, that is, $x_k \in (c_A)'$. Therefore, by Lemma 5.3, there exist numbers α_k and series $\sum_l \eta_{kl}$ satisfying the condition $\sum_l |\eta_{kl}| < \infty$ for every k, so that

$$x_k = \mu \alpha_k + \sum_l \eta_{kl} z_l; \; z_l := A_l x, \mu := \lim_l z_l. \tag{5.6}$$

Taking now $z_l = \delta_{lr}$ in (5.6), we see that $x_k = \eta_{kr}$, since $\mu = \lim_l \delta_{lr} = 0$. If $z_l = \delta_{ll}$ in (5.6), then

$$x_k = \alpha_k + \sum_l \eta_{kl},$$

since in that case $\mu = \lim_l \delta_{ll} = 1$. Hence the sequences (η_{kr}) and (η_k), where

$$\eta_k = \alpha_k + \sum_l \eta_{kl},$$

are solutions of the system of equations $z_l = A_l x$ $(l = 0, 1, \cdots)$ corresponding to $z_l = \delta_{lr}$ and $z_l = \delta_{ll}$. Thus, from (5.6), we get (5.5).

Remark 5.1 If A is a normal method, then $A^{-1} := (\eta_{kl})$ is the inverse matrix of A, and, between η_k and A^{-1}, the relationship

$$\eta_k = \sum_{l=0}^{k} \eta_{kl} \tag{5.7}$$

holds (of course, in this case, A^{-1} is also normal).

Example 5.1 For the Riesz method $A = (\tilde{R}, p_n)$ (or $A = (\overline{N}, p_n)$) relation (5.5) for every $x = (x_k) \in c_{(\tilde{R},p_n)}$ takes the form

$$x_k = \frac{P_k}{p_k} z_k - \frac{P_{k-1}}{p_k} z_{k-1}, z_k := (\tilde{R}, p_n)_k x,$$

by Remark 5.1, since $(\tilde{R}, p_n)$ is normal with inverse matrix $A^{-1} = (\eta_{kl})$, where (see [16], p. 115)

$$\eta_{kl} = \begin{cases} P_k/p_k & (l = k), \\ -P_{k-1}/p_k & (l = k - 1), \\ 0 & (l < k - 1 \text{ or } l > k). \end{cases}$$

Definition 5.5 A matrix A is called sequence-to-sequence conservative (or Sq–Sq conservative) if $c \subset c_A$, and sequence-to-sequence regular (shortly, Sq–Sq regular) if $A \in (c, c; P)$ (see Section 1.2), that is,

$$\lim_n A_n x = \lim_n x_n$$

for each $x \in c$.

Definition 5.6 A method A is called series-to-sequence conservative (or Sr–Sq conservative) if $cs \subset c_A$, and series-to-sequence regular (or Sr–Sq regular) if

$$\lim_n A_n x = \lim_n X_n$$

for each $x = (x_k) \in cs$.

Definition 5.7 A method A is called series-to-series conservative (or Sr–Sr conservative) if $cs \in cs_A$, and series-to-series regular (or Sr–Sr regular) if

$$\lim S(Ax) = \lim Sx$$

for each $x = (x_k) \in cs$.

Definition 5.8 A method A is called series-to-sequence absolutely conservative (or Sr–Sr absolutely conservative) if $l \in bv_A$.

For all sequences $x := (x_k)$ and $y := (y_k)$ we set, correspondingly,

$$X_n := \sum_{k=0}^{n} x_k \text{ and } Y_n := \sum_{k=0}^{n} y_k.$$

Let $A = (a_{nk})$, $\tilde{A} = (\tilde{a}_{nk})$, $\hat{A} = (\hat{a}_{nk})$ and $\overline{A} = (\overline{a}_{nk})$ be matrices with real or complex entries and

$$Y_n = \sum_k a_{nk} x_k, \; Y_n = \sum_k \tilde{a}_{nk} X_k, \; y_n = \sum_k \hat{a}_{nk} x_k, \; y_n = \sum_k \overline{a}_{nk} X_k.$$

Then we say that A, $\tilde{A}$, $\hat{A}$, and $\overline{A}$ determine, correspondingly, the series-to-sequence, the sequence-to-sequence, series-to-series, and sequence-to-series transforms. If matrices A, $\tilde{A}$, $\hat{A}$, and $\overline{A}$ are triangular, then all of the transforms are related (see [16], pp. 50–53).

Example 5.2 Let $A = (a_{nk})$ and $\tilde{A} = (\tilde{a}_{nk})$ be triangular matrices. We show that

$$a_{nk} = \sum_{l=k}^{n} \tilde{a}_{nl}, \tag{5.8}$$

$$\tilde{a}_{nk} = \Delta_k a_{nk}. \tag{5.9}$$

Indeed, the validity of (5.8) follows from the relation

$$Y_n = \sum_{l=0}^{n} \tilde{a}_{nl} X_l = \sum_{l=0}^{n} \tilde{a}_{nl} \sum_{k=0}^{l} x_k = \sum_{k=0}^{n} \left(\sum_{l=k}^{n} \tilde{a}_{nl} \right) x_k.$$

Now, using (5.8), we obtain

$$\Delta_k a_{nk} = \sum_{l=k}^{n} \tilde{a}_{nl} - \sum_{l=k+1}^{n} \tilde{a}_{nl} = \tilde{a}_{nk},$$

that is, relation (5.9) holds.

Using the relationships between A, $\tilde{A}$, $\hat{A}$, and $\overline{A}$, we can assert, for example, that, if A is an Sr–Sq conservative method, then $\tilde{A}$ and $\hat{A}$ are, correspondingly, Sq–Sq conservative and Sr–Sr conservative methods. Similarly, if $\tilde{A}$ is Sq–Sq regular, then A and $\hat{A}$ are, correspondingly, Sr–Sq regular and Sr–Sr regular methods.

5.3 The Existence Conditions of Matrix Transform *Mx*

In this section, we give necessary and sufficient conditions for the existence of a matrix transform Mx for every $x \in c_A$, or $x \in bv_A$, where M is an arbitrary matrix and A is a matrix method. We see that series $M_n x$ are convergent for every $x \in c_A$ or $x \in bv_A$ if and only if the numbers m_{nk}, for every fixed n, are convergence factors, correspondingly, for c_A and bv_A. We first find necessary and sufficient conditions for the numbers ε_k to be convergence factors, correspondingly, for c_A and bv_A. For a reversible method A, we use the so-called inverse transformation method, which was developed by Kangro (see [31]), for the problem of summability factors.

Let

$$d_{jl} := \sum_{k=0}^{j} \varepsilon_k \eta_{kl}; \quad \sum_l |\eta_{kl}| < \infty,$$

where (ε_k) is a sequence of numbers, and (η_{kr}), for each r, is the solution of the system of equations $z_l = A_l x$ $(l = 0, 1, \cdots)$ for $z_l = \delta_{lr}$, where $\delta_{lr} = 0$ for $r \neq l$ and $\delta_{ll} = 1$.

Lemma 5.5 Let $A = (a_{nk})$ be a reversible method. Numbers ε_k are convergence factors for c_A if and only if

there exist the finite limits $\lim_j d_{jl} := d_l$, $\qquad (5.10)$

$(\varepsilon_k \eta_k) \in cs$, $\qquad (5.11)$

$\sum_l |d_{jl}| = O(1)$. $\qquad (5.12)$

Proof: **Necessity.** We assume that the numbers ε_k are convergence factors for c_A, that is, $(\varepsilon_k x_k) \in cs$ for every $x = (x_k) \in c_A$, and show that conditions (5.10)–(5.12) are satisfied. As every element of $x = (x_k) \in c_A$ can be represented by (5.5), we have

$$\sum_{k=0}^{j} \varepsilon_k x_k = \mu \sum_k \varepsilon_k \eta_k + \sum_l d_{jl}(z_l - \mu); \quad z_l := A_l x, \mu := \lim_l z_l, \qquad (5.13)$$

since the change of order of summation in the second summand in the right side of this equation is allowed by the relation

$$\sum_l |\eta_{kl}(z_l - \mu)| = O(1) \sum_l |\eta_{kl}| < \infty.$$

It is not difficult to see that condition (5.11) holds, since $(\eta_k) \in c_A$. Hence it follows from (5.13) that the finite limit

$$\lim_j \sum_l d_{jl}(z_l - \mu)$$

exists for the sequence $(z_l - \mu) \in c_0$. Conversely, as A is a reversible method, for each $z = (z_l) \in c_0$, there exists a sequence $x = (x_k) \in c_A$ so that $z_l = A_l x$ and

$\mu = 0$. Therefore, the matrix $D := (d_{jl}) \in (c_0, c)$ exists. Thus, using Exercise 1.3, we obtain that conditions (5.10) and (5.12) are satisfied.

Sufficiency. We assume that conditions (5.10)–(5.12) are satisfied. Similar to the proof of the necessity condition of the present lemma, we can assert that equalities (5.13) hold for every $x = (x_k) \in c_A$. Conditions (5.10) and (5.12) imply (see Exercise 1.3) that $D \in (c_0, c)$. As $(z_l - \mu) \in c_0$ for every $x = (x_k) \in c_A$ with $\mu = \lim_l z_l$, where $z_l = A_l x$, then $(\varepsilon_k x_k) \in cs$ for every $x = (x_k) \in c_A$ by condition (5.11).

Let

$$h^n_{jl} := \sum_{k=0}^{j} m_{nk} \eta_{kl}.$$

From Lemma 5.5 and Exercise 5.4, we immediately get the following results.

Proposition 5.1   Let $A = (a_{nk})$ be a reversible method, and $M = (m_{nk})$ an arbitrary matrix. The matrix transformation $y = Mx$ exists for each $x \in c_A$ if and only if

there exist finite limits $\lim_j h^n_{jl} := h_{nl}$, (5.14)

$(m_{nk}\eta_k) \in cs$ for every fixed n, (5.15)

$\sum_l |h^n_{jl}| = O_n(1)$. (5.16)

Proposition 5.2   Let $A = (a_{nk})$ be a reversible method, and $M = (m_{nk})$ an arbitrary matrix. The matrix transformation $y = Mx$ exists for each $x \in bv_A$ it and only if conditions (5.14) and (5.15) are fulfilled and

$$\sum_{l=0}^{r} h^n_{jl} = O_n(1).$$ (5.17)

In the special case in which the reversible method $A = (a_{nk})$ is normal, A has an inverse matrix $A^{-1} = (\eta_{nk})$. Therefore, for a normal A, $D := (d_{jl})$ and $H^n := (h^n_{jl})$ for all n are lower-triangular matrices with

$$d_{jl} := \sum_{k=l}^{j} \varepsilon_k \eta_{kl}, \quad h^n_{jl} := \sum_{k=l}^{j} m_{nk} \eta_{kl}.$$

Lemma 5.6   Let $A = (a_{nk})$ be a normal method. The numbers ε_k are convergence factors for cs_A if and only if condition (5.10) is satisfied, and

$$\sum_l |\Delta_l d_{jl}| = O(1).$$ (5.18)

Moreover, for every $x := (x_k) \in cs_A$, the equalities

$$\sum_k \varepsilon_k x_k = d_0 \lim Sy + \sum_l \Delta_l d_l (Y_l - \lim Sy) \qquad (5.19)$$

are satisfied, with

$$Y_l := \sum_{k=0}^l y_k, \qquad (5.20)$$

where $y := (y_k)$ and $y_k = A_k x$.

Proof: **Necessity.** Let numbers ε_k be convergence factors for cs_A, that is, $(\varepsilon_k x_k) \in cs$ for every $x = (x_k) \in c_A$. We show that conditions (5.10) and (5.18)–(5.19) with (5.20) are fulfilled. For each $x \in cs_A$, we can write

$$\sum_{k=l}^j \varepsilon_k x_k = \sum_{l=0}^j d_{jl} y_l = D_j y,$$

where $y = (y_l) \in cs$. As method A is normal, then, for every $y \in cs$, there exists an $x \in cs_A$ such that $Ax = y$. Hence $D = (d_{jl}) \in (cs, c)$, and

$$\lim_j D_j y = M_n x \qquad (5.21)$$

for every $x \in cs_A$, where $y = Ax$. This implies that conditions (5.10) and (5.18) are satisfied and equality (5.19) with (5.20) are true by Theorem 1.5.

Sufficiency. If conditions (5.10) and (5.18) are satisfied, then, by Theorem 1.5, we have $D \in (cs, c)$. Since equations (5.21) hold, then equalities (5.19) with (5.20) also hold by Theorem 1.5. □

From Lemma 5.6, we immediately get the following result.

Proposition 5.3 Let $A = (a_{nk})$ be a normal method and $M = (m_{nk})$ an arbitrary matrix. The matrix transformation $y = Mx$ exists for each $x \in cs_A$ if and only if condition (5.14) is fulfilled and

$$\sum_{l=0}^j |\Delta_l h_{jl}^n| = O_n(1). \qquad (5.22)$$

Moreover, for every $x := (x_k) \in cs_A$, the equations

$$M_n x = h_0 \lim Sy + \sum_l \Delta_l h_{nl}(Y_l - \lim Sy) \qquad (5.23)$$

are satisfied with (5.20), where $y := (y_k) = (A_k x)$.

Proposition 5.3 first was proved in [11].

5.4   Matrix Transforms for Reversible Methods

We describe necessary and sufficient conditions for $M \in (c_A, c_B)$, $M \in (bv_A, c_B)$, and $M \in (bv_A, bv_B)$, where $A = (a_{nk})$ is a reversible method, $B = (b_{nk})$ a lower triangular method, and $M = (m_{nk})$ is an arbitrary matrix. As in Section 5.3, by (η_{kr}) and (η_k), we denote the solutions of the system of equations $z_l = A_l x$ $(l = 0, 1, \dots)$, correspondingly, for $z_l = \delta_{lr}$ $(r = 0, 1, \cdots)$ and $z_l = \delta_{ll}$. For an arbitrary B, we also present sufficient conditions. Let $G = (g_{nk}) = BM$, that is,

$$g_{nk} := \sum_{l=0}^{n} b_{nl} m_{lk}$$

and

$$\gamma_{nl}^r = \sum_{k=0}^{r} g_{nk} \eta_{kl}.$$

Theorem 5.1   Let $A = (a_{nk})$ be a reversible method, $B = (b_{nk})$ a lower triangular method, and $M = (m_{nk})$ an arbitrary matrix. Then, $M \in (c_A, c_B)$ if and only if conditions (5.14)–(5.16) are fulfilled and

there exists the finite limit $\lim_{n} \sum_{k} g_{nk} \eta_k := \gamma$,   (5.24)

there exist the finite limits $\lim_{n} \gamma_{nl} := \gamma_l$,   (5.25)

$$\sum_{l} |\gamma_{nl}| = O(1),$$ (5.26)

where

$$\gamma_{nl} := \lim_{r} \gamma_{nl}^r.$$

Proof: **Necessity.** We assume that $M \in (c_A, c_B)$. Then, by Proposition 5.1, conditions (5.14)–(5.16) are satisfied and

$$B_n y = G_n x$$ (5.27)

for every $x := (x_k) \in c_A$, where $y = (y_k) = (M_k x)$. This implies that $c_A \subset c_G$. Hence condition (5.24) is fulfilled, since $\eta \in c_A$. As every element x_k of a sequence $x := (x_k) \in c_A$ may be presented in the form (5.5), then we have

$$\sum_{k=0}^{r} g_{nk} x_k = \mu \sum_{k=0}^{r} g_{nk} \eta_k + \sum_{l} \gamma_{nl}^r (z_l - \mu),$$ (5.28)

because

$$\sum_{l} |\eta_{kl}| |z_l - \mu| = O(1) \sum_{l} |\eta_{kl}| < \infty.$$

It now follows from (5.28) that the finite limits

$$\lim_r \sum_l \gamma_{nl}^r (z_l - \mu) \tag{5.29}$$

exist for the sequence $(z_l - \mu) \in c_0$, because the series $G_n x$ converges for this $x \in c_A$. From the other side, by the reversibility of A for every $z := (z_l) \in c_0$, there exists a sequence $x := (x_k) \in c_A$ such that $z_l = A_l x$ and $\mu = 0$. Consequently, the matrix $\Gamma^n := (\gamma_{nl}^r) \in (c_0, c)$ for every n. Hence, from (5.28), we have, using Exercise 1.3, that

$$\sum_k g_{nk} x_k = \mu \sum_{k=0} g_{nk} \eta_k + \sum_l \gamma_{nl}(z_l - \mu) \tag{5.30}$$

for every $x = (x_k) \in c_A$. It is not difficult to see that $(z_l - \mu) \in c_\Gamma$, where $\Gamma := (\gamma_{nl})$. Moreover, $\Gamma \in (c_0, c)$ by the reversibility of A. Therefore, due to Exercise 1.3, conditions (5.25) and (5.26) are satisfied.

Sufficiency. Let conditions (5.14)–(5.16) and (5.24)–(5.26) be fulfilled. Then, the matrix transformation $y = Mx$ exists for every $x \in c_A$ by Proposition 5.1. Therefore, equations (5.27) hold for every $x \in c_A$, where $y = (y_k) = (M_k x)$, and the series $G_n x$ converge for every $x \in c_A$. Consequently (see the proof of the necessity part), for every $x \in c_A$, equalities (5.28) and (5.30), where $z_k = A_k x$ and $\mu = \lim_k z_k$, hold. Moreover, the sequence $(z_l - \mu) \in c_\Gamma$ by (5.25) and (5.26). Hence, $x \in c_\Gamma$ by (5.24) if $x \in c_A$. From (5.27), it follows now that $y = (y_k) \in c_B$. Thus, $M \in (c_A, c_B)$.

Remark 5.2 For an infinite matrix B, equation (5.27) generally does not hold for each $x \in c_A$, and limits (5.27) do not exist for the sequence $(z_l - \mu)$, that is, Γ^n does not belong into (c_0, c), in general. Therefore, for formulating necessary and sufficient conditions for $M \in (c_A, c_B)$, we need to assume the validity of (5.27) for each $x \in c_A$, and add necessary and sufficient conditions for $\Gamma^n \in (c_0, c)$ to the conditions of Theorem 5.1. Thus, for a nontriangular method B, we have the following: if A is a reversible method, M an arbitrary matrix, and equalities (5.27) hold for each $x \in c_A$, then $M \in (c_A, c_B)$ if and only if conditions (5.14)–(5.16), (5.24)–(5.26) are satisfied, and

the finite limit $\lim_r \gamma_{nk}^r := \gamma_{nk}$ exists, $\tag{5.31}$

$$\sum_k |\gamma_{nk}^r| = O_n(1), \tag{5.32}$$

where

$$g_{nk} := \sum_l b_{nl} m_{lk}.$$

From Theorem 5.1, we immediately get the following result.

Corollary 5.1   Let A be a reversible method, B a lower-triangular method, and M an arbitrary matrix. If $M \in (c_A, c_B)$, then $\sum_l |\gamma_l| < \infty$.

Since it follows from $M \in (c_A, c_B)$ that $c_A \subset c_G$ (see the proof of Theorem 5.1), then, from Theorem 5.1, we get the following corollary.

Corollary 5.2   If A is a reversible Sr–Sq or Sq–Sq conservative method, B a lower-triangular method, and M an arbitrary matrix belonging to (c_A, c_B), then G is correspondingly, an Sr–Sq or an Sq–Sq conservative matrix method.

If the elements m_{nk} of a matrix M are presented by equality (5.1), where (ε_k) is a sequence of numbers, then the transformation $y = Mx$ always exists, and equations (5.27) hold for every $x \in c_A$. Therefore, in this case, conditions (5.14)–(5.16) are fulfilled, and conditions (5.24)–(5.26) and (5.31)–(5.32) are necessary and sufficient for the numbers (ε_k) to be the summability factors of type (A, B) for a reversible method A, and an arbitrary matrix method B (this result for first proved in [31]). We note that the numbers (ε_k) are said to be summability factors of type (A, B), if $(\varepsilon_k x_k) \in c_B$ for every $x := (x_k) \in c_A$. If, in addition, $\varepsilon_k \equiv 1$, then conditions (5.24)–(5.26) and (5.31)–(5.32) give us the Theorem of Mazur–Hill, determining necessary and sufficient conditions for $c_A \subset c_B$ (see, e.g., [16], p. 64).

Now we consider the problem of M^{seq}-consistency of matrix methods A and B on c_A.

Theorem 5.2   Let $A = (a_{nk})$ be a reversible method, $B = (b_{nk})$ a lower triangular method, and $M = (m_{nk})$ an arbitrary matrix. The methods A and B are M^{seq}-consistent on c_A if and only if conditions (5.14)–(5.16) and (5.24)–(5.26), with $\gamma_k \equiv 0$ and $\gamma = 1$, are fulfilled.

Proof: **Necessity.** We assume that A and B are M^{seq}-consistent on c_A, that is, $c_A \in c_G$ and

$$\lim_n G_n x = \lim_n A_n x \qquad (5.33)$$

for every $x \in c_A$. Then, $M \in (c_A, c_B)$, and hence conditions (5.14)–(5.16) and (5.24)–(5.26) are fulfilled. Moreover, $\gamma = 1$, since $\lim_n A_n \eta = 1$ by the definition of η. From the proof of Theorem 5.1, we see that equations (5.28) hold for every $x \in c_A$, where $z_k = A_k x$ and $\mu = \lim_k z_k$. This implies

$$\lim_n \sum_l \gamma_{nl}(z_l - \mu) = 0$$

for every $x \in c_A$. Consequently, $\Gamma \in (c_0, c_0)$ by the reversibility of A. Therefore, due to Exercise 1.4, $\gamma_k \equiv 0$.

Sufficiency. Let conditions (5.14)–(5.16) and (5.24)–(5.26) with $\gamma_k \equiv 0$ and $\gamma = 1$ be fulfilled. Then, equations (5.28) hold for every $x \in c_A$, where $z_k = A_k x$ and $\mu = \lim_k z_k$. Thus, using Exercise 1.4, $\Gamma \in (c_0, c_0)$. Therefore, it follows from (5.28) that equality (5.33) holds for every $x \in c_A$, since $\gamma = 1$. This means that A and B are M^{seq}-consistent on c_A.

The proof of necessary and sufficient conditions for the matrix transforms from bv_A into c_B or into bv_B, and from c_A into bv_B, is similar to the proof of these conditions for the transformation from c_A into c_B. Therefore, we leave the proofs of these conditions to the reader (see Exercises 5.5–5.7).

Remark 5.3 In the case of a nontriangular method B, similarly to Remark 4.1 for formulating necessary and sufficient conditions for $M \in (bv_A, c_B)$, $M \in (bv_A, bv_B)$ or $M \in (c_A, bv_B)$, we need to assume the validity of (5.27) for each $x \in c_A$, or $x \in bv_A$, and add necessary and sufficient conditions for $\Gamma^n \in (bv_0, c)$, or $\Gamma^n \in (c_0, c)$ to the corresponding results. We advise the reader to formulate the above-mentioned results. If the elements m_{nk} of a matrix M are represented by equality (5.1), where (ε_k) is a sequence of numbers, then, from these results, it is possible to get necessary and sufficient conditions for (ε_k) to be the summability factors of type $(|A|, B)$, $(|A|, |B|)$ or $(|A|, B)$ for a reversible method A and an arbitrary matrix method B, introduced in [31]. We note that the numbers (ε_k) are said to be the summability factors of type $(|A|, B)$, (type $(|A|, |B|)$ if $(\varepsilon_k x_k) \in c_B$ $((\varepsilon_k x_k) \in bv_B$, respectively) for every $x := (x_k) \in bv_A$. The numbers (ε_k) are said to be the summability factors of type $(A, |B|)$, if $(\varepsilon_k x_k) \in bv_B$ for every $x := (x_k) \in c_A$.

For a nontriangular method B, we now give sufficient conditions for $M \in (c_A, c_B)$, $M \in (bv_A, c_B)$, $M \in (bv_A, bv_B)$, or $M \in (c_A, c_B)$. To accomplish this, we first find necessary and sufficient conditions for the validity of (5.27) for each $x \in c_A$ or $x \in bv_A$. To find these conditions, we use the following result.

Lemma 5.7 (see [38], pp. 257–258). Let $M = (m_{nk})$ be an arbitrary matrix. The series

$$\sum_k \left(\sum_n u_n m_{nk} \right) x_k \tag{5.34}$$

converges for every $(u_n) \in l$ if and only if

$$m_{nk} = O_k(1) \tag{5.35}$$

and $x := (x_k) \in b_M$, where

$$b_M := \left\{ x = (x_k) \ : \ \sum_{k=l}^{\infty} m_{nk} x_k = O(1) \right\}.$$

In addition, if series (5.34) converges for every $(u_n) \in l$, then

$$\sum_k \left(\sum_n u_n m_{nk} \right) x_k = \sum_n u_n M_n x.$$

With the help of Lemma 5.7, we immediately get the following example.

Example 5.3 Let $A = (a_{nk})$, $M = (m_{nk})$ be arbitrary matrices, and $B = (b_{nk})$ a matrix satisfying the condition

$$\sum_k |b_{nk}| = O_n(1). \tag{5.36}$$

Equality in (5.27) holds for each $x \in c_A$ $(x \in bv_A)$ if and only if condition (5.35) is fulfilled and $c_A \subset b_M$ $(bv_A \subset b_M$, respectively).

Now we prove the following propositions.

Proposition 5.4 Let $A = (a_{nk})$ be a reversible method and $M = (m_{nk})$ an arbitrary matrix. Then, $c_A \subset b_M$ if and only if conditions (5.14)–(5.15) are fulfilled and

$$\sum_{k=0}^r m_{nk} \eta_k = O(1), \tag{5.37}$$

$$\sum_l |h_{jl}^n| = OO(1). \tag{5.38}$$

Proof: **Necessity.** Let $c_A \subset b_M$. Then, the transformation $y = Mx$ exists for each $x \in c_A$ and $\eta \in b_M$, since $\eta \in c_A$. Hence, conditions (5.14), (5.15), and (5.37) are satisfied.

By Lemma 5.4, the elements x_k of the sequence $x := (x_k) \in c_A$ may be represented by (5.5). This implies that the series

$$\sum_l \eta_{kl} (z_l - \mu)$$

are convergent. Consequently, the equality

$$\sum_{k=0}^j m_{nk} x_k = \sum_{k=0}^j m_{nk} \eta_k + \sum_l h_{jl}^n (z_l - \mu) \tag{5.39}$$

holds for each $x \in c_A$. Therefore,

$$\sum_l h_{jl}^n (z_l - \mu) = O(1)$$

for each $(z_l - \mu) \in c_0$ by (5.37) since A is a reversible method. Moreover, the H_j^n, defined with the help of the equalities

$$H_j^n(z) := \sum_l h_{jl}^n z_l$$

for each $z := (z_k) \in c_0$, are continuous linear functionals on c_0. Hence, by the principle of uniform boundedness, we get that the sequence of the norms of $H_j^n \in (c_0)'$ is uniformly bounded. Therefore, condition (5.38) is fulfilled.

Sufficiency. We now assume that conditions (5.14), (5.15), (5.37), and (5.38) are fulfilled. Then, we see that condition (5.16) is also satisfied. Consequently, the transformation $y = Mx$ exists for each $x \in c_A$ by Proposition 5.1. Further, equalities (5.39) are valid for each $x := (x_k) \in c_A, \eta \in b_M$ by (5.37), and

$$|h_j^n(z_l - \mu)| \le \sum_l |h_{jl}^n||z_l - \mu| = O(1)$$

by (5.38) (since $(z_l - \mu) \in c_0$). This implies $c_A \subset b_M$.

With the help of Example 5.3, Proposition 5.4, and Exercise 5.8, we can find sufficient conditions for $M \in (c_A, c_B)$, $M \in (bv_A, c_B)$, $M \in (bv_A, bv_B)$, or $M \in (c_A, c_B)$, if B is a matrix method satisfying condition (5.36).

Theorem 5.3   Let $A = (a_{nk})$ be a reversible method, and $B = (b_{nk}), M = (m_{nk})$ matrices, satisfying, correspondingly, conditions (5.36) and (5.35). If conditions (5.14), (5.15), (5.37), and (5.38) are fulfilled, then condition (5.31) holds. If, in addition to it, conditions (5.24)–(5.26) are fulfilled, then $M \in (c_A, c_B)$.

Proof: Equality (5.27) holds for each $x \in c_A$ by Example 5.3 and Proposition 5.4. Hence, it is sufficient to show that $c_A \subset c_G$. With the help of Lemma 5.4, we obtain that the elements x_k of the sequence $x := (x_k) \in c_A$ may be represented as (5.5), then equalities (5.28) are true for every $x \in c_A$. Now with the help of conditions (5.35), (5.36), and (5.38), we obtain that

$$\sum_r |b_{nr}| \sum_{k=0}^{j} |m_{rk} \eta_{kl}| = O_{j,l}(1) \sum_r |b_{nr}| = O_{j,l,n}(1),$$

that is,

$$\sum_r |b_{nr}| \sum_{k=0}^{j} |m_{rk} \eta_{kl}| = O_{j,l,n}(1), \tag{5.40}$$

and

$$\sum_r |b_{nr} h_{jl}^r| = O_n(1) \tag{5.41}$$

by condition (5.38). This implies that

$$\gamma_{nl}^j = \sum_r b_{nr} h_{jl}^r, \tag{5.42}$$

and condition (5.31) is satisfied by (5.14). Moreover, condition (5.15) is fulfilled because

$$\sum_r |b_{nr}| \sum_l h_{jl}^r = O_n(1)$$

by (5.38). Consequently, from (5.28) we get, due to Exercises 1.1 and 1.3, and condition (5.38), that equalities (5.30) are true for every $x \in c_A$. Besides, using Exercise 1.3, we conclude that conditions (5.24) and (5.25) imply the existence of the finite limits

$$\lim_n \sum_l \gamma_{nl}(z_l - \mu) \tag{5.43}$$

for every $x \in c_A$, because $(z_l - \mu) \in c_0$ for every $x \in c_A$. Thus, $c_A \subset c_G$ by (5.38), and therefore $M \in (c_A, c_B)$.

Theorem 5.4   Let $A = (a_{nk})$ be a reversible method, and $B = (b_{nk})$, $M = (m_{nk})$ matrices satisfying, correspondingly, conditions (5.36) and (5.35). If conditions (5.14), (5.15), (5.37) and

$$\sum_{l=0}^k h_{jl}^n = O(1) \tag{5.44}$$

are fulfilled, then condition (5.31) is fulfilled. If, in addition, conditions (5.24), (5.25) and

$$\sum_{k=0}^l \gamma_{nk} = O(1) \tag{5.45}$$

are fulfilled, then $M \in (bv_A, c_B)$.

Proof: Equality (5.27) holds for each $x \in c_A$ by Example 5.3 and Exercise 5.8. Hence, it is sufficient to show that $bv_A \subset c_G$. It is not difficult to see that equalities (5.28) hold for every $x \in bv_A$ since $bv_A \subset c_A$. As

$$|h_{jl}^n| = \left| \sum_{k=0}^l h_{jk}^n - \sum_{k=0}^{l-1} h_{jk}^n \right| = O(1)$$

by condition (5.44), then relations (5.41) and (5.42) hold. Consequently, condition (5.31) is satisfied by condition (5.14). In addition,

$$\sum_r |b_{nr}| \sum_{k=0}^l \sum_{i=0}^j m_{ri} \eta_{ik} = O_{j,l}(1) \sum_r |b_{nr}| = O_{j,l,n}(1)$$

by (5.35) and (5.36). This implies that

$$\sum_{l=0}^{k} \gamma_{nl}^{j} = \sum_{r} b_{nr} \sum_{l=0}^{k} h_{jl}^{r}.$$

Thus, condition

$$\sum_{l=0}^{k} \gamma_{nl}^{j} = O_n(1)$$

is fulfilled by (5.36) and (5.44). Hence, from (5.28) it follows that (5.30) holds for every $x \in bv_A$ by Theorem 1.6 and condition (5.24) is satisfied. Therefore, conditions (5.25) and (5.45) imply by Theorem 1.6 that there exist the finite limits (5.43) for each $x \in bv_A$. Consequently, $bv_A \subset c_G$ by (5.24). Thus, $M \in (bv_A, c_B)$.

From the proofs of Theorems 5.3 and 5.4, we can formulate the following remark.

Remark 5.4 If a matrix $B = (b_{nk})$ satisfies the condition

$$\sum_{k} |b_{nk}| = O(1), \tag{5.46}$$

then condition (5.26) is redundant in Theorem 5.3, and condition (5.45) is redundant in Theorem 5.4.

We note that Theorem 5.1 first was proved in [1] and Theorems 5.3–5.4 in [2].

5.5 Matrix Transforms for Normal Methods

In this section, we describe necessary and sufficient conditions for $M \in (cs_A, cs_B)$, where $A = (a_{nk})$ is a normal method, $B = (b_{nk})$ is a lower-triangular method, and $M = (m_{nk})$ is an arbitrary matrix. Also we consider the inclusions $l_A \subset l_B$ and $l_A \subset cs_B$. In the proofs, we use the inverse transformation method.

Theorem 5.5 Let $A = (a_{nk})$ be a normal method, $B = (b_{nk})$ a lower triangular method and $M = (m_{nk})$ an arbitrary matrix. Then, $M \in (cs_A, cs_B)$ if and only if conditions (5.14) and (5.22) are fulfilled, and

$$\text{there exist the finite limits} \lim_{r} \sum_{n=0}^{r} \gamma_{nl} := \widehat{\gamma}_l, \tag{5.47}$$

$$\sum_{l} \left| \sum_{n=0}^{r} \Delta_l \gamma_{nl} \right| = O(1). \tag{5.48}$$

Proof: **Necessity.** Let $M \in (cs_A, cs_B)$. Then, the matrix transformation $y = Mx$ exists for each $x \in cs_A$. Hence conditions (5.14) and (5.22) are satisfied, and equalities (5.23) with (5.20) hold for every $x \in cs_A$ by Proposition 5.3. This implies that equality

$$B_n(Mx) = \gamma_{n0} \lim Sy + \sum_l \Delta_l \gamma_{nl}(Y_l - \lim Sy) \tag{5.49}$$

holds for every $x \in cs_A$, where $y = Ax$. As A is a normal method, then, for $e^0 \in cs$, there exists a sequence $\tilde{x} \in cs_A$ such that $A\tilde{x} = e^0$. Therefore, due to $\tilde{x} = ((A^{-1})_k e^0)$, we get

$$B_n(M\tilde{x}) = B_n[M(A^{-1}e^0)] = \gamma_{n0}.$$

Consequently,

$$\text{the series } \sum_n \gamma_{n0} \text{ is convergent.} \tag{5.50}$$

As every $Y = (Y_l) \in c$ may be represented in the form (5.20), where $y = (y_k) \in cs$, and, for this y, there exists an $x \in cs_A$ such that $Ax = y$. Since A is normal, then, from (5.49) and (5.50), we can conclude that series

$$\sum_n \sum_l \Delta_l \gamma_{nl}(Y_l - \lim Sy) \tag{5.51}$$

converges for every $Y = (Y_l) \in c$. As every $Y = (Y_l) \in c$ may be represented in the form

$$Y = Y^0 + e \lim Sy; \quad Y^0 = (Y_k^0) \in c_0,$$

series (5.51) converges for each $Y^0 = (Y_k - \lim Sy) \in c_0$, that is, $\Gamma := (\Delta_l \gamma_{nl}) \in (c_0, cs)$. Hence, condition (5.48) is satisfied, and the series

$$\sum_n \Delta_l \gamma_{nl}$$

converges for all l by Theorem 1.7. Consequently, condition (5.47) is satisfied by (5.50).

Sufficiency. We assume that conditions (5.14), (5.22), (5.47), and (5.48) are satisfied. Then, the transformation $y = Mx$ exists for each $x \in cs_A$, and equalities (5.23) are true for each $x \in cs_A$ by Proposition 5.3. Hence, equalities (5.49) also hold. Now from (5.47) and (5.48), we get $\Gamma \in (c_0, cs)$. It follows from (5.49) that $M \in (cs_A, cs_B)$ by (5.47).

Theorem 5.6 Let $A = (a_{nk})$ be a normal method, $B = (b_{nk})$ a lower triangular method, and $M = (m_{nk})$ an arbitrary matrix. Then, A and B are M^{ser}-consistent if and only if conditions (5.14), (5.22), (5.48), and (5.47) with $\hat{\gamma}_l \equiv 1$ are satisfied.

Proof: **Necessity.** Assuming that A and B are M^{ser}-consistent, we get that conditions (5.14), (5.22), and (5.48) are satisfied by Theorem 5.5, and equalities (5.49) hold for each $x \in cs_A$. This implies that

$$\lim S[B(Mx)] = \lim S(Ax) \qquad (5.52)$$

for every $x \in cs_A$. Let $\tilde{x} \in cs_A$ be a sequence satisfying the equality $A\tilde{x} = e^0$. Then, $\lim S(A\tilde{x}) = 1$, and we have

$$\lim S[B(M\tilde{x})] = \hat{\gamma}_0 = 1. \qquad (5.53)$$

Hence it follows from (5.49) and (5.52) that $\Gamma \in (c_0, cs_0)$. Therefore

$$\sum_n \Delta_l \gamma_{nl} = 0$$

for all l by Theorem 1.9. Thus, $\hat{\gamma}_l \equiv 1$ by (5.53).

Sufficiency. Suppose that all of the conditions of Theorem 5.6 are satisfied. Then, we have $M \in (cs_A, cs_B)$ by Theorem 5.5 and equalities (5.49) are valid for every $x \in cs_A$. This implies that $\Gamma \in (c_0, cs_0)$ by Theorem 1.8. Hence, from (5.49), we get, with the help of (5.48), that equality (5.52) holds for each $x \in cs_A$, that is, A and B are M^{ser}-consistent.

We consider now the case if $M = (m_{nk})$ is a lower triangular factorable matrix, that is,

$$m_{nk} = t_n u_k, k \leq n,$$

where (t_n) and (u_k) are sequences of numbers. Let F be the set of all lower triangular factorable matrices M. We start with simple examples.

Example 5.4 Let $A = (a_{nk})$ be a method with $e^0 \in c_A$, $B = (b_{nk})$ an arbitrary method and $M \in$ F. If $M \in (c_A, c_B)$, then $(t_n) \in c_B$, since

$$M_n e^0 = t_n u_0.$$

Example 5.5 Let $A = (a_{nk})$ and $B = (b_{nk})$ be arbitrary methods and $B^t = (b^t_{pn})$ be a matrix, defined by the relation $b^t_{pn} = b_{pn} t_n$. Then, it is easy to verify that $M \in$ F belongs to (c_A, c_B) if

$$(u_k x_k) \in cs \text{ for every } x \in c_A, \qquad (5.54)$$

$$B^t \text{ is Sq–Sq conservative.} \qquad (5.55)$$

Proposition 5.5 Let $B = (b_{nk})$ be an Sr–Sq regular method, where $b_{nk} > 0$ for all n and k, and (t_n) is a sequence of numbers. Then, condition (5.55) is satisfied if and only if $(t_n) \in l$.

Proof: **Necessity.** Let B^t be Sq–Sq conservative. We show that $(t_n) \in l$. First we see, due to Exercise 1.1, that

$$S_r := \sum_n |b_{rn} t_n| = \sum_n b_{rn} |t_n| = O(1). \tag{5.56}$$

If $\sum_n |t_n| = \infty$, then (see [25], p. 92) $\lim_{r \to \infty} S_r = \infty$, that is, condition (5.56) is not valid. This implies (see Exercise 1.1) that $(t_n) \in l$.

Sufficiency. Let $(t_n) \in l$. We see that $(t_n) \in c_B$, due to the Sr–Sq regularity of B. The Sr–Sq regularity of B also implies that $b_{nk} = O(1)$, and the finite limits $\lim_n b_{nk}$ exist by Theorem 1.4. Hence

$$S_r = O(1) \sum_n |t_n'| = O(1).$$

Therefore, using Exercise 1.1, we can conclude that B^t is Sq–Sq conservative.

Example 5.6 Let $A = (a_{nk})$, $B = (b_{nk})$ be arbitrary matrices and (t_n), (u_k) sequences of numbers, $l \subset c_B$, $(t_n) \in l$ and $M = (t_n u_k) \in F$. We show that then $M \in (c_A, c_B)$ if condition (5.54) is satisfied. First define

$$V_n := \sum_{k=0}^{n} u_k x_k$$

for every $x \in c_A$. Then, with the help of (5.54), we obtain that $(V_n) \in c$ for every $x \in c_A$. This implies that $(V_n) \in m$ for each $x \in c_A$. Consequently,

$$\sum_n |M_n x| = \sum_n |t_n V_n| = O(1) \sum_n |t_n| = O(1)$$

for every $x \in c_A$. Hence $M \in (c_A, c_B)$ because $l \subset c_B$.

Now we consider the inclusions $l_A \subset l_B$ and $l_A \subset cs_B$ in the special case, when $B = (b_{nk}) \in F$, that is, $b_{nk} = t_n u_k, k \leq n$. Let

$$F_u^{cs} := \{B \in F \mid (t_n) \in cs\}, \quad F_u^l := \{B \in F \mid (t_n) \in l\}$$

for a given sequence $u = (u_k)$. We need to find necessary and sufficient conditions for $l_A \subset cs_B$ (for $l_A \subset l_B$) for every $B \in F_u^{cs}$ (for every $B \in F_u^l$, respectively). We begin with the following necessary conditions.

Proposition 5.6 Let $A = (a_{nk})$ be a method such that $e^0 \in l_A$. Then, the following assertions are true:

1. If $l_A \subset cs_B$ for $B \in F$, then $(t_n) \in cs$.
2. If $l_A \subset l_B$ for $B \in F$, then $(t_n) \in l$.

Proof: The proof follows from the equality $B_n e^0 = t_n u_0$.

Theorem 5.7 Let $A = (a_{nk})$ be a normal method with inverse matrix $A^{-1} = (\eta_{nk})$. Then, $l_A \subset l_B$ for each $B \in \mathrm{F}_u^l$ if and only if

$$\sum_{k=l}^{m} u_k \eta_{kl} = O(1). \tag{5.57}$$

Proof: For every $x = (x_k) \in l_A$, we can write

$$x_k = \sum_{l=0}^{k} \eta_{kl} z_l,$$

where $z_l = A_l x$, since the inverse matrix A^{-1} of a normal matrix A is lower triangular. This implies that, for $B \in \mathrm{F}$ and for each $x \in l_A$, the relation

$$B_n x = t_n L_n(z); \quad L_n(z) := \sum_{l=0}^{n} \left(\sum_{k=l}^{n} u_k \eta_{kl} \right) z_l \tag{5.58}$$

is satisfied. As A is normal, for every $z = (z_l) \in l$ there exists an $x \in l_A$ such that $A_l x = z_l$. Hence from (5.58), we can conclude that $Bx \in l$ for every $B \in \mathrm{F}_u^l$ and every $x \in l_A$ if and only if $(t_n L_n(z)) \in l$ for all $(t_n) \in l$ and $z \in l$. This relation is true if and only if

$$L_n(z) = O_z(1) \tag{5.59}$$

for every $z \in l$. As

$$L_n(z) = \sum_{l=0}^{n} s_{nl} z_l, \quad \text{where } s_{nl} := \sum_{k=l}^{n} u_k \eta_{kl},$$

for each $z \in l$, then, for the validity of (5.59) for each $z \in l$, it is necessary and sufficient that $S := (s_{nk}) \in (l, m)$. Using Exercise 1.10, we get that $S \in (l, m)$ if and only if condition (5.57) is satisfied.

Theorem 5.8 Let $A = (a_{nk})$ be a normal method with inverse matrix $A^{-1} = (\eta_{nk})$. Then, $l_A \subset cs_B$ for each $B \in \mathrm{F}_u^{cs}$ if and only if

$$\sum_{k=l}^{\infty} |u_k \eta_{kl}| = O(1). \tag{5.60}$$

Proof: Similar to the proof of Theorem 5.7, for $B \in \mathrm{F}$ we get equality (5.58) for each $x \in l_A$, where $z_l = A_l x$. Hence, from (5.58), we obtain that $Bx \in cs$ for every $B \in \mathrm{F}_u^{cs}$ and $x \in l_A$ if and only if $(t_n L_n(z)) \in cs$ for each $(t_n) \in cs$ and each $z \in l$.

This last relation holds, by the well-known theorem of Dedekind–Hadamard, if and only if

$$\sum_n |\Delta_n L_n(z)| = O_z(1) \tag{5.61}$$

for each $z \in l$. In addition, for each $z \in l$ we can write

$$\Delta_n L_n(z) = -v_{n+1} \sum_{l=0}^{n+1} \eta_{n+1,l} z_l \text{ or } \Delta_n L_n(z) = \sum_{l=0}^{n+1} c_{n+1,l} z_l, \, c_{nl} := -u_n \eta_{nl}.$$

Consequently, (5.61) holds for each $z \in l$ if and only if $C := (c_{nl}) \in (l, l)$. From Exercise 1.11, we obtain that $C \in (l, l)$ if and only if condition (5.60) is satisfied.

It is easy to see that condition (5.57) follows from condition (5.60). Hence, from Theorems 5.7 and 5.8, we immediately get the following corollary.

Corollary 5.3 Let $A = (a_{nk})$ be a normal method and $u = (u_k)$ an arbitrary sequence. If $l_A \subset cs_B$ for each $B \in F_u^{cs}$, then $l_A \subset l_B$ for each $B \in F_u^l$.

From Theorems 5.7 and 5.8, we immediately get the following corollary.

Corollary 5.4 Let $A = (a_{nk})$ be a normal method and $u = (u_k)$ an arbitrary sequence. If $l_A \subset cs_B$ for each $B \in F_u^{cs}$ or $l_A \subset l_B$ for each $B \in F_u^l$, then

$$u_l \eta_{ll} = O(1). \tag{5.62}$$

We note that Theorems 5.5–5.6 are first proved in [11] and Theorems 5.7 and 5.8 in [10].

5.6 Excercise

Exercise 5.1   Which form does (5.5) take for the Cesàro method $A = (C^\alpha)$?

Exercise 5.2   How it is possible to present the general form of a continuous linear functional in c_0^l and $(c_A^0)'$?

Exercise 5.3   Prove that, for triangular A, $\tilde{A}$, $\hat{A}$, and $\overline{A}$, the following relations hold:

$$\overline{a}_{nk} = \tilde{a}_{nk} \tilde{a}_{n-1,k}; \; \hat{a}_{nk} = a_{nk} a_{n-1,k}; \; \tilde{a}_{nk} = \sum_{l=k}^n \tilde{a}_{lk}.$$

Exercise 5.4   Let $A = (a_{nk})$ be a reversible method. Prove that numbers ε_k are convergence factors for bv_A if and only if conditions (5.10) and (5.11) hold and

$$\sum_{l=0}^{r} d_{jl} = O(1).$$

Hint. The proof is similar to Lemma 5.5. Instead of Exercise 1.3, it is necessary to use Theorem 1.6.

Exercise 5.5   Let $A = (a_{nk})$ be a reversible method, $B = (b_{nk})$ a lower triangular method, and $M = (m_{nk})$ an arbitrary matrix. Prove that $M \in (bv_A, c_B)$ if and only if conditions (5.14), (5.15), (5.17), (5.24), (5.25), and (5.45) are fulfilled. Prove that, if $M \in (bv_A, c_B)$, then

$$\sum_{k=0}^{l} \gamma_k = O(1).$$

Hint. The proof is similar to Theorem 5.1. Relations (5.28) and (5.30) hold for every $x \in bv_A$ with $(z_l - \mu) \in bv_0$ and $\mu = \lim_l z_l$, and $\eta \in bv_A$. Use Theorem 1.6.

Exercise 5.6   Let $A = (a_{nk})$ be a reversible method, $B = (b_{nk})$ a lower triangular method, and $M = (m_{nk})$ an arbitrary matrix. Prove that $M \in (bv_A, bv_B)$ if and only if conditions (5.14), (5.15), and (5.17) are fulfilled, and

$$\eta \in bv_G, \tag{5.63}$$

$$\sum_{n=1}^{\infty} \left| \sum_{k=0}^{l} (\gamma_{nk} - \gamma_{n-1,k}) \right| + \left| \sum_{k=0}^{l} \gamma_{0k} \right| = O(1). \tag{5.64}$$

Hint. The proof is similar to Theorem 5.1. Use Exercise 1.13.

Exercise 5.7   Let $A = (a_{nk})$ be a reversible method, $B = (b_{nk})$ a lower triangular method, and $M = (m_{nk})$ an arbitrary matrix. Then, $M \in (c_A, bv_B)$ if and only if conditions (5.14)–(5.16) and (5.63) are fulfilled and

$$\sum_{n \in K} \sum_{k \in L} (\gamma_{nk} - \gamma_{n-1,k}) = O(1), \tag{5.65}$$

where K and L are arbitrary finite subsets of $\mathbf{N}$.

Hint. The proof is similar to Theorem 5.1. Use Theorem 1.9.

Exercise 5.8   Let $A = (a_{nk})$ be a reversible method and $M = (m_{nk})$ an arbitrary matrix. Prove that $bv_A \subset b_M$ if and only if conditions (5.14), (5.15), (5.37), and (5.44) are satisfied.

Hint. The proof is similar to that of Proposition 5.4.

Exercise 5.9   Let $A = (a_{nk})$ be a reversible method, and $B = (b_{nk})$, $M = (m_{nk})$ matrices, satisfying, correspondingly, conditions (5.36) and (5.35). Prove that if conditions (5.14), (5.15), (5.37), and (5.44) are fulfilled, then condition (5.31) is fulfilled. Prove that if, in addition, conditions (5.63) and (5.64) are satisfied, then $M \in (bv_A, bv_B)$.
Hint. The proof of this result is similar to those of Theorems 5.3 and 5.4. Use Theorem 1.9 and Exercise 1.13.

Exercise 5.10   Let $A = (a_{nk})$ be a reversible method, and $B = (b_{nk})$, $M = (m_{nk})$ matrices, satisfying, correspondingly, conditions (5.36) and (5.35). Prove that if conditions (5.14), (5.19), (5.37), and (5.38) are fulfilled, then condition (5.31) is fulfilled. Prove that if, in addition, conditions (5.63) and (5.65) are satisfied, then $M \in (c_A, bv_B)$.
Hint. See the hint of Exercise 5.9.

Exercise 5.11   Let $A = (a_{nk})$ be an Sr–Sr-conservative normal method, $B = (b_{nk})$ a lower triangular method and $M = (m_{nk})$ an arbitrary matrix. Prove that, if $M \in (cs_A, cs_B)$, then

$$\sum_n g_{nk} = \widehat{g}_k \ (\widehat{g}_k \text{ is a finite number}). \tag{5.66}$$

Hint. Use Theorem 5.5.

Exercise 5.12   Let A be an Sr–Sr-regular normal method, $B = (b_{nk})$ a lower triangular method, and $M = (m_{nk})$ an arbitrary matrix. Prove that, if A and B are M^{ser}-consistent on cs_A, then condition (5.66) holds with $\widehat{g}_k = 1$
Hint. Use Theorem 5.6.

References

1 Aasma, A.: Preobrazovanija polei summirujemosti (Matrix transformations of summability fields). Tartu Riikl. Ül. Toimetised **770**, 38–50 (1987).
2 Aasma, A.: Characterization of matrix transformations of summability fields. Tartu Ül. Toimetised **928**, 3–14 (1991).
3 Aasma, A.: Matrix transformations of summability fields of normal regular matrix methods. Tallinna Tehnikaül. Toimetised. Matem. Füüs. **2**, 3–10 (1994).
4 Aasma, A.: Matrix transformations of summability fields of regular perfect matrix methods. Tartu Ül. Toimetised **970**, 3–12 (1994).
5 Aasma, A.: On the matrix transformations of absolute summability fields of reversible matrices. Acta Math. Hung. **64**(2), 143–150 (1994).

6 Aasma, A.: Matrix transformations of summability domains of generalized matrix methods in Banach spaces. Rend. Circ. Mat. Palermo (2) **58**(3), 467–476 (2009).

7 Aasma, A.: Some notes on matrix transforms of summability domains of Cesàro matrices. Math. Model. Anal. **15**(2), 153–160 (2010).

8 Aasma, A.: Factorable matrix transforms of summability domains of Cesàro matrices. Int. J. Contemp. Math. Sci. **6**(41–44), 2201–2206 (2011).

9 Aasma, A.: Some classes of matrix transforms of summability domains of normal matrices. Filomat **26**(5), 1023–1028 (2012).

10 Aasma, A.: Some inclusion theorems for absolute summability. Appl. Math. Lett. **25**(3), 404–407 (2012).

11 Aasma, A.: Matrix transforms of summability domains of normal series-to-series matrices. J. Adv. Appl. Comput. Math. **1**, 35–39 (2014).

12 Ahmad, Z.U. and Khan, F.M.: Absolute Nörlund summability factors of infinite series with applications. Indian J. Math. **16**(3), 137–156 (1974).

13 Alpár, L.: Sur certains changements de variable des séries de Faber (Certain changes of variables in Faber series). Stud. Sci. Math. Hung. **13**(1-2), 173–180 (1978).

14 Alpár, L.: Cesàro Summability and Conformal Mapping. Functions, Series, Operators, Vols **I, II**. (Budapest, 1980), 101–125. Colloq. Math. Soc. János Bolyai, Vol. **35**. North-Holland, Amsterdam (1983).

15 Alpár, L.: On the linear transformations of series summable in the sense of Cesàro. Acta Math. Hung. **39**(1-2), 233–243 (1982).

16 Baron, S.: Vvedenie v teoriyu summiruemosti ryadov (Introduction to the Theory of Summability of Series). Valgus, Tallinn (1977).

17 Baron, S. and Kiesel, R.: Absolute ϕ-summability factors with a power for A^{α}-methods. Analysis **15**(4), 311–324 (1995).

18 Baron, S. and Peyerimhoff, A.: Complete proofs of the main theorems on summability factors. Acta Comment. Univ. Tartu. Math. **3**, 31–61 (1999).

19 Boos, J.: Classical and Modern Methods in Summability. Oxford University Press, Oxford (2000).

20 Bor, H. and Leindler, L.: A new note on absolute summability factors. Rend. Circ. Mat. Palermo (2) **160**(1–2), 75–81 (2011).

21 Bor, H.; Yu, D. and Zhou, P.: Some new factor theorems for generalized absolute Cesàro summability. Positivity **19**(1), 111–120 (2015).

22 Bosanquet, L.S.: Convergence and summability factors in a sequence. Mathematika **1**, 24–44 (1954).

23 Bosanquet, L.S.: Convergence and summability factors in a sequence. II. Mathematika **30**(2), 255–273 (1983).

24 Butzer, P.L. and Nessel, R.I.: Fourier Analysis and Approximation: One-Dimensional Theory. Birkhäuser Verlag, Basel and Stuttgart (1971).

25 Cooke, R.G.: Beskonetsnŏe matricŏ i prostranstva posledovatelnostei (Infinite Matrices and Sequence Spaces). State Publishing House of Physics-Mathematics Literature, Moscow (1960).

26 Dikshit, G.D. and Rhoades, B.E.: An inclusion relation between Cesàro and Nörlund matrices for absolute summability. J. Math. Anal. Appl. **170**(1), 171–195 (1992).

27 Hardy, G.H.: Divergent Series. Oxford University Press, Oxford (1949).

28 Jakimovski, A.; Russell, D.C. and Tzimbalario, J.: Inclusion theorems for matrix transformations. J. Anal. Math. **26**, 391–404 (1973).

29 Jakimovski, A.; Russell, D.C. and Tzimbalario, J.: Inclusion relations for general Riesz typical means. Can. Math. Bull. **17**, 51–61 (1974).

30 Jurkat, W.: Summierbarkeitsfaktoren. Math. Z. **58**, 186–203 (1953).

31 Kangro, G.: O množitelyah summirujemosti (On summability factors). Tartu. Gos. Univ. Trudy Estest.-Mat. Fak. **37**, 191–229 (1955).

32 Kangro, G. and Vichmann, F.: Abstraktnye mnnožiteli summiruyemosti dlya metoda vzvechennyh srednyh Riesza (Abstract summability factors for the method of weighted Riesz means). Tartu Riikl. Ül. Toimetised **102**, 209–225 (1961).

33 Leiger, T.: Množitely summiruyemosty dlya obobchennyh metodov summirovanya (Summability factors for generalized summability methods). Tartu Riikl. Ül. Toimetised **504**, 58–73 (1981).

34 Leiger, T.: Vklyutchenye obobchennyh metodov summirovanya (Inclusion of generalized summability methods). Tartu Riikl. Ül. Toimetised **504**, 17–34 (1981).

35 Leiger, T.: Funktsionaalanalüüsi meetodid summeeruvusteoorias (Methods of functional analysis in summability theory). Tartu Ülikool, Tartu (1992).

36 Peyerimhoff, A.: Konvergenz- und Summierbarkeitsfaktoren. Math. Z. **55**, 23–54 (1951).

37 Peyerimhoff, A.: Untersuchungen über absolute Summierbarkeit. Math. Z. **57**, 265–290 (1953).

38 Russell, D.C.: Inclusion theorems for section-bounded matrix transformations. Math. Z. **113**, 255–265 (1970).

39 Russell, D.C.: Note on convergence factors. II. Indian J. Math. **13**(1), 29–44 (1971).

40 Thorpe, B.: Matrix transformations of Cesàro summable series. Acta Math. Hung. **48**, 255–265 (1986).

41 Trebels, W.: Multipliers for (C, α)-bounded Fourier Expansions in Banach Spaces and Approximation Theory, Lecture Notes in Mathematics, vol. **329**. Springer-Verlag, Berlin-Heidelberg, New York (1973).

42 Vichmann, F.: Rashirenie metoda Peyerimhoffa dlya slutchaya obobchennyh množitelei summiruyemosti (An extension of the method of Peyerimhoff to the case of generalized summability factors). Tartu Riikl. Ül. Toimetised **129**, 170–193 (1962).

43 Vichmann, F.: Obobchennye mnnožiteli summiruyemosti dlya metoda vzvechennyh srednyh Riesza (Generalized summability factors for the weighted means method of Riesz). Tartu Riikl. Ül. Toimetised **129**, 199–224 (1962).

44 Wilansky, A.: Summability through Functional Analysis, North-Holland Mathematics Studies, Vol. 85; Notas de Matemática (Mathematical Notes), Vol. 91. North-Holland Publishing Co., Amsterdam (1984).

45 Zhuk, V.V.: Approksimatsiya periodicheskikh funktsii (Approximation of periodic functions). Leningrad State University, Leningrad (1982).

6

Matrix Transformations of Summability and Absolute Summability Domains: Peyerimhoff's Method

6.1 Introduction

First, we note that all of the notions and notations not defined in this chapter can be found in Chapters 1 and 5. In this chapter, we continue to study necessary and sufficient conditions for M to be transformed from c_A into c_B, and the M-consistency of matrix methods A and B on c_A started in Chapter 5. The notions of perfect matrix methods and AK-spaces are introduced and some properties of these methods and these spaces are presented. We consider the case when A is a regular perfect method and B a triangular method. Separately, the cases, if c_A^0 is a BK-AK-space, are studied. We note that the first results for this case were found by A. Aasma in 1994 (see [1, 2]). For a proof of results of [1, 2], he used a functional analytic method, worked out by Peyerimhoff for studying the problem of summability factors (see [6]). Also, this method has been used in the papers [4, 7–10], and in monographs [3, 5, 11, 12].

6.2 Perfect Matrix Methods

We begin with the following notions.

Definition 6.1 An Sr-Sq regular method A is called perfect if Δ is a fundamental set for c_A.

Definition 6.2 An Sq-Sq regular method A is called perfect if $\Delta \cup e$ is a fundamental set for c_A.

Lemma 6.1 If A is an Sq-Sq regular perfect method, then every $x = (x_k) \in c_A$ may be represented in the form

$$x = x^0 + \xi e, \text{ where } x^0 \in c_A^0, \xi = \lim_k A_k x. \tag{6.1}$$

An Introductory Course in Summability Theory, First Edition. Ants Aasma, Hemen Dutta, and P.N. Natarajan.
© 2017 John Wiley & Sons, Inc. Published 2017 by John Wiley & Sons, Inc.

Proof: Relation (6.1) for $x \in c_A$ follows from the relation $x - \xi e \in c_A^0$ for each $x = (x_k) \in c_A$ if $\xi = \lim_k A_k x$.

It is clear that an Sq-Sq (Sr-Sq) conservative method A is perfect if and only if $c_A = \overline{c}$ (correspondingly, $c_A = \overline{cs}$; the notation $\overline{X}$ means the closure of a space X). In this section, we consider only Sq-Sq conservative methods. For the characterization of perfect methods, we need some additional notations and notions. Let

$$\varphi := \{x = (x_k) \in \omega : \text{there exists a } k_0 \in N \text{ so that } x_k = 0 \text{ for } k > k_0\rangle$$

and *kern f* is the kernel of $f \in (c_A)'$, that is,

$$kern\, f := \{x \in c_A : f(x) = 0\}.$$

Definition 6.3 A functional $f \in (c_A)'$ is said to be a test function if

$$kern\, f \supset \varphi, \tag{6.2}$$

and there exists a presentation of f in form (5.3) with $\alpha = 0$. $\tag{6.3}$

In general, the representation (5.3) of $f \in (c_A)'$ is not unique. However, there exists a great class of matrices A, for which the α in the representation of f does not depend on the different presentations of this functional in the form (5.3).

Definition 6.4 A matrix A is called α-unique if all presentations for every $f \in (c_A)'$ in form (5.3), have the same value $\alpha := \alpha(f)$.

For a conservative method $A = (a_{nk})$, we define

$$\rho(A) := \delta - \sum_k \delta_k,$$

where

$$\delta_k := \lim_n A_n e^k \quad \text{and} \quad \delta := \lim_n A_n e.$$

Definition 6.5 An Sq-Sq conservative method A is called coregular if $\rho(A) \neq 0$, and conull if $\rho(A) = 0$.

Example 6.1 Every Sq-Sq regular method is coregular. Indeed, in this case $\rho(A) = 1 \neq 0$ due to $\delta_k \equiv 0$ and $\delta = 1 \neq 0$ (see Theorem 1.1).

Example 6.2 The method $A_{(-1,1)}$, introduced in Section 1.1, is conull, because, in that case, $\rho(A) = 0$, due to $\delta_k \equiv 0$ and $\delta = 0$.

Example 6.3   We show that every coregular matrix is α-unique. For this purpose we suppose, by contradiction, that a coregular matrix A is not α-unique. Then, there exist at least two different values α_1 and α_2, such that

$$f(x) = \alpha_1 \lim_n A_n x + \sum_n t_n A_n x, \ (t_n) \in l, \tag{6.4}$$

$$f(x) = \alpha_2 \lim_n A_n x + \sum_n t'_n A_n x. \tag{6.5}$$

Subtracting (6.4) from (6.5), we obtain

$$0 = (\alpha_2 - \alpha_1) \lim_n A_n x + \sum_n (t'_n - t_n) A_n x. \tag{6.6}$$

Now, for $x = e^k$ and $x = e$, from (6.6), we, correspondingly, get

$$0 = (\alpha_2 - \alpha_1)\delta_k + \sum_n (t'_n - t_n) a_{nk} \tag{6.7}$$

and

$$0 = (\alpha_2 - \alpha_1)\delta + \sum_n (t'_n - t_n) \sum_k a_{nk}. \tag{6.8}$$

Summing k from 0 to ∞ in (6.7), we obtain

$$0 = (\alpha_2 - \alpha_1) \sum_k \delta_k + \sum_n (t'_n - t_n) \sum_k a_{nk}. \tag{6.9}$$

The change in the order of summation of the second summand of (6.9) is allowed by the Sq-Sq conservativity of A (then condition (1.1) holds) and the relation $(t'_n - t_n) \in l$. Comparing (6.8) and (6.9), we get $\delta - \sum_k \delta_k$ or $\rho(A) - 0$ if $\alpha_1 \neq \alpha_2$. It means that A is not coregular, which gives us the contradiction. Hence every coregular matrix is α-unique.

Now we present an example of an $f \in (c_A)'$ satisfying condition (6.3).

Example 6.4   Let A be an Sq-Sq conservative method and $f \in (c_A)'$ such a functional that $\operatorname{kern} f \supset \varphi$. We show that the representation (5.3) of an f has $\alpha = 0$ if and only if $f(e) = 0$. Indeed, a coregular method A is α-unique by Example 6.3. Let

$$\gamma(f) := f(e) - \sum_k f(e^k). \tag{6.10}$$

For every $f \in (c_A)'$, the relation

$$f(x) = \alpha v_A(x) + \sum_k x_k f(e^k); \ \ v_A(x) := \lim_n A_n x - \sum_k \delta_k x_k$$

holds for all $x \in c_A$ if the series $\sum_k x_k f(e^k)$ is convergent (see [3], p. 411, or [5], pp. 64–65). Then, from (6.10) we obtain

$$\gamma(f) = \alpha v_A(e) = \alpha \gamma(A), \tag{6.11}$$

where we consider A to be a continuous linear operator on c_A. Our assertion follows from (6.11). It means that condition (6.3) is equivalent to the condition $f(e) = 0$.

Let

$$T_A := \{t = (t_n) \in l : \quad \text{for every } x \in c_A \text{ the series } \sum_k \left(\sum_n t_n a_{nk} \right) x_k$$

converges$\}$.

Example 6.5 We prove that $f \in (c_A)'$ is a test function if and only if it can be represented in the form

$$fx = \sum_n t_n A_n x - \sum_k \left(\sum_n t_n n_{nk} \right) x_k \ (t \in T_A). \tag{6.12}$$

Indeed, it is not difficult to see that a function f, defined by (6.12), belongs $(c_A)'$, $\text{kern } f \supset \varphi$ and $\alpha(f) = 0$. Hence f is a test function for every $t \in T_A$.

Conversely, for every test function $f \in (c_A)'$ there exists $t \in l$ such that

$$fx = \sum_n t_n A_n x (x \in c_A),$$

with

$$0 = f e^k = \sum_n t_n a_{nk}.$$

Hence $t \in T_A$ and f can be represented by (6.12).

Let

$$P_A := \{x = (x_k) \in c_A : \text{for every } t \in T_A : \sum_n t_n A_n x = \sum_k \left(\sum_n t_n a_{nk} \right) x_k \}.$$

Lemma 6.2 (see [3], p. 444, or [5], p. 85) For a coregular method A the relation $P_A = \bar{c} = c_A$ holds.

Using Example 6.5, we obtain

$$P_A := \{x = (x_k) \in c_A : f(x) = 0 \text{ for every test function } f\}.$$

Definition 6.6 A matrix A is said to be of type M if

$$t \in l \text{ and } \sum_n t_n a_{nk} = 0 \text{ imply } t_n = 0. \tag{6.13}$$

It is easy to see that (6.13) holds if and only if the system of equations

$$\sum_n t_n a_{nk} = 0 \ (k \in N) \tag{6.14}$$

has only the zero-solution in l.

Theorem 6.1   A reversible coregular method A is perfect if and only if it is of type M.

Proof: As $f \in (c_A)'$ can be presented in form (5.3), then f is a test function if and only if

$$f(x) = \sum_n t_n A_n x \text{ and } kern\, f \supset \varphi,$$

that is, (6.14) holds. Conversely, the perfectness of A implies $P_A = c_A$ by Lemma 6.2. This is possible only in the case that the zero-functional is the unique test function, that is, relation (6.13) holds.   □

Example 6.6   We prove that a normal coregular method A is perfect if the columns of the inverse matrix $A^{-1} := (\eta_{kl})$ of A are bounded sequences. Indeed, the columns of A^{-1} are A-summable, since

$$\sum_{k=l}^{n} a_{nk} \eta_{kl} = \delta_{nl}.$$

As $t \in l$ and condition (1.1) holds for a conservative method due to Exercise 1.1, then

$$t_l = \sum_{n=l}^{\infty} \left(\sum_{k=l}^{n} a_{nk} \eta_{kl} \right) t_n = \sum_{k=l}^{\infty} \eta_{kl} \sum_{n=k}^{\infty} t_n a_{nk}.$$

This implies that if (6.14) holds, then $t_l \equiv 0$, that is, A is of type M. Hence A is perfect by Theorem 6.1.

Example 6.7   An Sq-Sq conservative Riezs method $(\tilde{R}, p_n)$ is perfect. Indeed, since each column of the inverse matrix of $(\tilde{R}, p_n)$ has only two elements, different from 0 (see Section 5.2), then the columns of the inverse matrix of $(\tilde{R}, p_n)$ are bounded sequences, and therefore, by Example 6.4, $(\tilde{R}, p_n)$ is perfect.

6.3   The Existence Conditions of Matrix Transform *Mx*

In this section, we give necessary and sufficient conditions for the existence of a matrix transform Mx for every $x \in c_A$, where M is an arbitrary matrix and A is a regular perfect method. For this purpose, we use a functional analytic method, which was developed for the problem of summability factors by Peyerimhoff (see [6]). We note that all notations, not mentioned here, are taken from Chapter 5.

Lemma 6.3 Let $A = (a_{nk})$ be an Sq-Sq regular perfect method such that c_A^0 is a BK-space. The numbers ε_k are the convergence factors for c_A if and only if there exist functionals $f_j \in (c_A^0)'$ so that

$$f_j(e^k) = \begin{cases} \varepsilon_k & (k \leq j), \\ 0 & (k > j), \end{cases} \tag{6.15}$$

and

$$\|f_j\|_{(c_A^0)'} = O(1). \tag{6.16}$$

Proof: **Necessity.** We assume that the numbers ε_k are the convergence factors for c_A. Let

$$f_j(x^0) = \sum_{k=0}^{j} \varepsilon_k x_k^0 \tag{6.17}$$

for every $x^0 := (x_k^0) \in c_A^0$. Then, $f_j \in (c_A^0)'$, and hence condition (6.15) is satisfied. Condition (6.16) is fulfilled by the principle of uniform boundedness, since c_A^0 is a BK-space, and the finite limit $\lim_j f_j(x^0)$ exists for every $x^0 \in c_A^0$.

Sufficiency. We suppose that all of the conditions of Lemma 6.3 are satisfied, and show that numbers ε_k are the convergence factors for c_A. First we prove that (6.17) holds for every $x^0 := (x_k^0) \in c_A^0$. For this purpose, let us denote

$$h_j(x^0) := f_j(x^0) - \sum_{k=0}^{j} \varepsilon_k x_k^0$$

for each $x^0 \in c_A^0$. It is easy to see that $h_j \in (c_A^0)'$ and, in addition, $h_j(e^k) = 0$ by (6.15). Consequently, $h_j(x^0) = 0$ on the fundamental set Δ of the space c_A^0. Therefore, $h_j(x^0) = 0$ for each $x^0 \in c_A^0$. Thus, (6.17) is valid for each $x^0 \in c_A^0$. As each $x \in c_A$ may be represented in the form (6.1) by Lemma 6.1, we get

$$\sum_{k=0}^{j} \varepsilon_k x_k = f_j(x^0) + \xi \sum_{k=0}^{j} \varepsilon_k \tag{6.18}$$

for every $x := (x_k) \in c_A$. Moreover, $\lim_j f_j(e^k) = \varepsilon_k$ by condition (6.15), that is, the sequence (f_j) converges on the fundamental set Δ of the space c_A^0. This implies, by condition (6.16) and the Theorem of Banach–Steinhaus, that the finite limit $\lim_j f_j(x^0)$ exists for each $x^0 \in c_A^0$. As $c_0 \subset c_A^0$, the finite limit $\lim_j f_j(x^0)$ exists for each $x^0 \in c_0$, that is, $(\varepsilon_k x_k^0) \in cs$ for each $x^0 \in c_0$ by (6.17). This implies that $(\varepsilon_k) \in l$, due to Exercise 1.3. Indeed, we may define a lower-triangular matrix

$C = (c_{nk})$ by $c_{nk} := \varepsilon_k$ if $k \leq n$, and $c_{nk} := 0$ if $k > n$. Then, $(\varepsilon_k x_k^0) \in cs$ for each $x^0 \in c_0$ if and only if $C \in (c_0, c)$. From this inclusion, using Exercise 1.3, we obtain that the validity of condition

$$\sum_{k=0}^{n} |c_{nk}| = O(1)$$

is necessary. Hence $(\varepsilon_k) \in l$, and also $(\varepsilon_k) \in cs$. Therefore, from (6.18), we conclude that $(\varepsilon_k x_k) \in cs$ for every $x = (x_k) \in c_A$. Thus, the numbers ε_k are the convergence factors for c_A. □

Using Lemma 6.3 and Exercise 6.2, we immediately have the following results:

Proposition 6.1 Let $A = (a_{nk})$ be an Sq-Sq regular perfect method such that c_A^0 is a BK-space, and $M = (m_{nk})$ is an arbitrary matrix. The matrix transformation $y = Mx$ exists for each $x \in c_A$ if and only if there exist functionals $f_{nj} \in (c_A^0)'$ so that

$$f_{nj}(e^k) = \begin{cases} m_{nk} & (k \leq j), \\ 0 & (k > j) \end{cases} \tag{6.19}$$

and

$$\|f_{nj}\|_{(c_A^0)'} = O_n(1). \tag{6.20}$$

Proposition 6.2 Let $A = (a_{nk})$ be an Sr-Sq regular perfect method such that c_A is a BK-space and $M = (m_{nk})$ is an arbitrary matrix. The matrix transformation $y = Mx$ exists for each $x \in c_A$ if and only if there exist functionals $f_{nj} \in (c_A)'$, so that condition (6.19) is fulfilled and

$$\|f_{nj}\|_{(c_A)'} = O_n(1). \tag{6.21}$$

Now we consider, as an example, the special case where A is an Sq-Sq regular reversible perfect method.

Example 6.8 Let $A = (a_{nk})$ be an Sq-Sq regular reversible perfect method. We prove that numbers ε_k are the convergence factors for c_A if and only if there exists series $\sum_r \tau_{jr}$, satisfying the condition

$$\sum_r |\tau_{jr}| = O(1), \tag{6.22}$$

such that

$$\sum_r \tau_{jr} a_{rk} = \begin{cases} \varepsilon_k & (k \leq j), \\ 0 & (k > j). \end{cases} \tag{6.23}$$

As a reversible method A is a BK-space by Lemma 5.3, then it is sufficient to show that, in this case, conditions (6.15) and (6.16) may be represented, correspondingly, as in (6.23) and (6.22). It is easy to see that the f_j, defined by (6.17) for each $x^0 := (x^0_k) \in c^0_A$, belongs to $(c^0_A)'$. As $\lim_r A_r x^0 = 0$ for each $x^0 \in c^0_A$, since A is Sq-Sq regular, by Lemma 5.3 there exists series $\sum_r \tau_{jr}$, satisfying the condition

$$\sum_r |\tau_{jr}| = O_j(1), \tag{6.24}$$

such that

$$f_j(x^0) = \sum_r \tau_{jr} A_r x^0 \tag{6.25}$$

for every $x^0 \in c^0_A$. As $e^k \in c^0_A$, from (6.25) we get

$$f_j(e^k) = \sum_{r=0}^{j} \tau_{jr} a_{rk}. \tag{6.26}$$

Hence, condition (6.15) may be represented in form (6.23).

Further, we show that

$$\|f_j\|_{c^0_A} = \sum_r |\tau_{jr}|. \tag{6.27}$$

Indeed, for $x = (x_k) \in c^0_A$, we have

$$|f_j x| = \left| \sum_r \tau_{jr} A_r x \right| \le \sum_r |\tau_{jr}| |A_r x| \le \|x\|_{c^0_A} \sum_r |\tau_{jr}| = O_j(1)$$

by (6.24), since $\|x\|_{c^0_A} = \sup_r |A_r x|$. Hence

$$\|f_j\|_{c^0_A} = \sup_{\|x\|_{c^0_A} \le 1} |f_j x| \le \sum_r |\tau_{jr}|. \tag{6.28}$$

Conversely, let a sequence $y^{n,j} := (y^{n,j}_r) \in c_0$ for every fixed j be defined by

$$y^{n,j}_r = \begin{cases} sgn \ \tau_{jr} & (r \le n), \\ 0 & (r > n). \end{cases} \tag{6.29}$$

As A is a reversible method, then there exists a sequence $x^{n,j} \in c^0_A$ such that $Ax^{n,j} = y^{n,j}$. As

$$\|x^{n,j}\|_{c^0_A} = \sup_r |A_r x^{n,j}| = \sup_r |y^{n,j}_r| = 1$$

for all j an n, then

$$f_j x^{n,j} = \sum_r \tau_{jr} sgn \ \tau_{jr} = \sum_r |\tau_{jr}|.$$

Consequently,

$$\|f_j\|_{c^0_A} = \sup_{\|x\|_{c^0_A} \le 1} |f_j x| \ge f_j x^{n,j} = \sum_r |\tau_{jr}|. \tag{6.30}$$

From (6.28) and (6.30), it follows that relation (6.27) holds. Therefore, condition (6.16) may be represented by (6.22).

Using Example 6.8 and Exercise 6.3, we immediately get the following results.

Proposition 6.3 Let $A = (a_{nk})$ be an Sq-Sq regular reversible perfect method and $M = (m_{nk})$ an arbitrary matrix. The matrix transformation $y = Mx$ exists for each $x \in c_A$ if and only if there exists series $\sum_r \tau_{jr}^n$, satisfying the condition

$$\sum_r |\tau_{jr}^n| = O_n(1), \tag{6.31}$$

such that

$$\sum_r \tau_{jr}^n a_{rk} = \begin{cases} m_{nk} & (k \le j), \\ 0 & (k > j). \end{cases} \tag{6.32}$$

Proposition 6.4 Let $A = (a_{nk})$ be an Sr-Sq regular reversible perfect method and $M = (m_{nk})$ an arbitrary matrix. The matrix transformation $y = Mx$ exists for each $x \in c_A$ if and only if there exists series $\sum_r \tau_{jr}^n$, satisfying condition (6.31), and a bounded sequence τ_j^n for every fixed n such that

$$\tau_j^n + \sum_r \tau_{jr}^n a_{rk} = \begin{cases} m_{nk} & (k \le j), \\ 0 & (k > j). \end{cases} \tag{6.33}$$

We note that Propositions 6.1 and 6.2 were first proved in [2].

6.4 Matrix Transforms for Regular Perfect Methods

In this section, we describe necessary and sufficient conditions for $M \in (c_A, c_B)$, where $A = (a_{nk})$ is a regular perfect method, $B = (b_{nk})$ is a lower triangular method, and $M = (m_{nk})$ is an arbitrary matrix.

Theorem 6.2 Let $A = (a_{nk})$ be an Sq-Sq regular perfect method such that c_A^0 is a BK-space, $B = (b_{nk})$ is a lower triangular method, and $M = (m_{nk})$ is an arbitrary matrix. Then, $M \in (c_A, c_B)$ if and only if

the finite limits $\lim_n g_{nk} := g_k$ exist, $\qquad\qquad (6.34)$

the finite limit $\lim_n \sum_k g_{nk} := g$ exists, $\qquad\qquad (6.35)$

and there exist functionals $f_{nj} \in (c_A^0)'$ so that conditions (6.19) and (6.20) are satisfied, and

$$\|F_n\|_{(c_A^0)'} = O(1), \tag{6.36}$$

where the functionals F_n are defined on c_A^0 by the equalities

$$F_n(x) = \sum_{r=0}^{n} b_{nr} f_r(x),$$ (6.37)

with

$$f_r(x) = \lim_j f_{rj}(x).$$ (6.38)

Proof: **Necessity.** We assume that $M \in (c_A, c_B)$. Then, the equations

$$\sum_{r=0}^{n} b_{nr} \sum_{k=0}^{\infty} m_{rk} x_k = \sum_k g_{nk} x_k = G_n x$$

hold for each $x = (x_k) \in c_A$. This implies that $c_A \subset c_G$. As A is Sq-Sq regular, the method G is Sq-Sq conservative. Hence, conditions (6.34) and (6.35) are satisfied, due to Exercise 1.1. As the transformation $y = Mx$ exists for each $x \in c_A$, then by Proposition 6.1, there exist functionals $f_{nj} \in (c_A^0)'$ so that conditions (6.19) and (6.20) are fulfilled. It is easy to see that these functionals may be represented on c_A^0 in the form

$$f_{nj}(x) = \sum_{k=0}^{j} m_{nk} x_k.$$ (6.39)

Hence

$$f_r(x) = \lim_j f_{rj}(x) = M_r x$$ (6.40)

for each $x \in c_A^0$. Then, $f_r \in (c_A^0)'$ and, consequently, the functionals F_n, defined on c_A^0 by (6.37), are continuous and linear on c_A^0. Moreover, $F_n(x) = G_n x$ for each $x \in c_A^0$. It is not difficult to see that the sequence of continuous linear functionals (F_n) is convergent on the Banach space c_A^0. This implies that condition (6.36) is satisfied by the principle of uniform boundedness.

Sufficiency. We assume that the conditions of Theorem 6.2 are satisfied and show that then $M \in (c_A, c_B)$. First we note that the transformation $y = Mx$ exists for each $x \in c_A$, and relation (6.36) is true on the fundamental set Δ of c_A^0, by Proposition 6.1. Consequently, (6.36) is true everywhere on c_A^0. This implies that relation (6.40) holds for each $x \in c_A^0$. Thus, $f_r \in (c_A^0)'$. Hence

$$f_r(e^k) = m_{rk}, F_n(e^k) = g_{nk}, F_n \in (c_A^0)',$$

and the equalities $F_n(x) = G_n x$ hold for each $x \in c_A^0$. Therefore, the sequence of continuous linear functionals (F_n) is convergent on the fundamental set Δ of c_A^0 by condition (6.34). By this reason, we get, with the help of (6.36) and the theorem of Banach–Steinhaus, that the finite limit $\lim_n F_n(x)$ exists for each

$x \in c_A^0$. As every $x \in c_A$ may be represented in the form (6.1) by Lemma 6.1, we can write

$$G_n x = F_n(x^0) + \xi \sum_k g_{nk} \tag{6.41}$$

for each $x \in c_A^0$, where $\xi := \lim A_n x$ and $x^0 \in c_A^0$. Consequently, there exists a finite limit $\lim_n G_n x$ for each $x \in c_A$ by (6.35). Thus, $M \in (c_A, c_B)$.

We now consider the special case when A is an Sq-Sq regular method such that c_A^0 is a BK-AK-space; it means that c_A^0 is simultaneously a BK-space and an AK-space.

Definition 6.7 A Banach sequence space X with a norm $\| \cdot \|$ is said to be an AK-space if $\Delta \subset X$, and convergence by sections exists in X, that is,

$$\lim_n \|x^{[n]} - x\| = 0$$

for each $x := (x_k) \in c_A^0$, where $x^{[n]} := (x_0, \ldots, x_n, 0, \ldots)$ is called a section of x for every n.

Definition 6.8 A Banach sequence space X with a norm $\| \cdot \|$ is said to be an SAK-space if $\Delta \subset X$, and weak convergence by sections exists in X, that is,

$$\lim_n |f(x^{[n]}) - f(x)| = 0$$

for each $x := (x_k) \in c_A^0$, and $f \in (c_A^0)'$.

Convergence by sections is equivalent to the weak convergence by sections, that is, the following result holds:

Lemma 6.4 (see [11], p. 176) A Banach sequence space X is an AK-space if and only if X is an SAK-space.

Lemma 6.5 If A is an Sq-Sq regular method such that c_A^0 is a BK-AK-space, then A is perfect.

Proof: It is not difficult to see that Δ is a fundamental set of c_A^0. Hence, with the help of Lemma 6.1, we can conclude that $\Delta \cup \{e\}$ is a fundamental set for c_A. Therefore, A is a perfect method.

Remark 6.1 We note that, for an Sq-Sq regular method A, the space c_A is not necessarily an AK-space (see, e.g., [11], pp. 214–215).

Theorem 6.3 Let $A = (a_{nk})$ be an Sq-Sq regular method such that c_A^0 is a BK-AK-space, $B = (b_{nk})$ is a normal method, and $M = (m_{nk})$ is an arbitrary matrix. Then, $M \in (c_A, c_B)$ if and only if conditions (6.34) and (6.35) are fulfilled, and there exist functionals $F_n \in (c_A^0)'$ so that condition (6.36) is fulfilled and

$$g_{nk} = F_n(e^k). \tag{6.42}$$

Proof: **Necessity.** We assume that $M \in (c_A, c_B)$. Then, all conditions of Theorem 6.2 are satisfied, since A is a perfect method, by Lemma 6.6. Now it is not difficult to conclude that F_n, defined by (6.37) and (6.38), belong to $(c_A^0)'$ and satisfy conditions (6.36) and (6.41).

Sufficiency. We assume that the conditions of Theorem 6.3 are satisfied, and show that $M \in (c_A, c_B)$. It is sufficient to show that the transformation $y = Mx$ exists for each $x \in c_A$ and $c_A \subset c_G$. First, we note that the equations

$$G_n x = \sum_k F_n(e^k) x_k = \lim_j F_n(e^{[j]}) = F_n x$$

hold for each $x = (x_k) \in c_A^0$ by (6.42), since c_A^0 is an AK-space. Hence the $G_n x$ exist for every $x \in c_A^0$ and n. In addition, by the normality of B, we can write

$$m_{nk} = \sum_{l=0}^{n} b_{nl}^{-1} g_{lk},$$

where (b_{nl}^{-1}) is the inverse matrix of B. This implies that

$$M_n x = \sum_{l=0}^{n} b_{nl}^{-1} G_l x$$

for every $x \in c_A^0$. Consequently, the transformation $y = Mx$ exists for each $x \in c_A^0$. Further, we see that the sequence $(m_{nk}) \in cs$ for every fixed n, since the equalities

$$\sum_k m_{nk} = \sum_{l=0}^{n} b_{nl}^{-1} \sum_{k=0}^{\infty} g_{lk}$$

hold by (6.35). Therefore, we have

$$M_n x = M_n(x^0) + \xi \sum_k m_{nk}$$

for every $x \in c_A^0$, where $\xi := \lim_n A_n x$ and $x^0 \in c_A^0$ since, by Lemma 6.1, equalities (6.1) hold. Hence, the transformation $y = Mx$ exists for each $x \in c_A$.

Further, the sequence (F_n) is convergent on the fundamental set Δ of c_A^0 by (6.34) and (6.42). Therefore, (F_n) is convergent on c_A^0 by (6.36) and the Banach–Steinhaus theorem. As equalities (6.41), where $\xi := \lim_n A_n x$ and $x^0 \in c_A^0$, hold for every $x \in c_A$, then $c_A \subset c_G$ by (6.35). Thus, we have $M \in (c_A, c_B)$.

Now we find necessary and sufficient conditions for the M^{seq}-consistency of methods A and B on c_A.

Theorem 6.4   Let $A = (a_{nk})$ be an Sq-Sq regular perfect method such that c_A^0 is a BK-space, $B = (b_{nk})$ is a lower triangular method and $M = (m_{nk})$ is an arbitrary matrix. Then, A and B are M^{seq}-consistent on c_A if and only if conditions (6.34) and (6.35), with $g_k \equiv 0$ and $g = 1$, are fulfilled, and there exist functionals $f_{nj} \in (c_A^0)'$, so that conditions (6.19), (6.20), and (6.36) hold.

Proof: **Necessity.** We suppose that A and B are M^{seq}-consistent on c_A. Then, obviously, $M \in (c_A, c_B)$, and therefore all conditions of Theorem 6.2 are satisfied. As A is an Sq-Sq regular method, then $\lim_n A_n e^k = 0$ and $\lim_n A_n e = 1$. This implies that $g_k \equiv 0$ and $g = 1$ by the M^{seq}-consistency of A and B on c_A.

Sufficiency. Assuming that all of the conditions of Theorem 6.4 are fulfilled, we conclude that $M \in (c_A, c_B)$ by Theorem 6.2. In addition, equalities (6.41), where $\xi := \lim_n A_n x$, $F_n(x^0) = G_n x^0$, and $x^0 \in c_A^0$, hold for every $x \in c_A$ (see the proof of Theorem 6.2). As the sequence of continuous linear functionals (F_n) is convergent on c_A^0, then its limit $F \in (c_A^0)'$, and $F(x) = 0$ on the fundamental set Δ of c_A^0 by $g_k \equiv 0$. This implies that $F(x) = 0$ for every $x \in c_A^0$. Hence A and B are M^{seq}-consistent on c_A by (6.41) since $g = 1$.

Now we consider, as an example, the special case, where A is a reversible method. Then, by Lemma 5.3, c_A^0 (c_A) is a BK-space if A is Sq-Sq regular (Sr-Sq regular, respectively).

Example 6.9   Let $A = (a_{nk})$ be an Sq-Sq regular reversible method, such that c_A^0 is an AK-space, $B = (b_{nk})$ a normal method, and $M = (m_{nk})$ an arbitrary matrix. We show that, in this case, $M \in (c_A, c_B)$ if and only if conditions (6.34) and (6.35) are fulfilled, and there exists series $\sum_r \tau_{jr}$, satisfying condition (6.22), such that

$$g_{jk} = \sum_r \tau_{jr} a_{rk}. \tag{6.43}$$

First, we show the necessity of the above-mentioned conditions. We assume that $M \in (c_A, c_B)$. Then, all of the conditions of Theorem 6.3 are fulfilled, since c_A^0 is a BK-space, by Lemma 5.3. If the F_j, defined by $F_j(x) = G_j x$ on c_A^0, belong to $(c_A^0)'$, then by Lemma 5.3, there exists series $\sum_r \tau_{jr}$, satisfying condition (6.22), such that

$$F_j(x) = \sum_r \tau_{jr} A_r x \tag{6.44}$$

for every $x \in c_A^0$. Moreover, similar to the proof of relation (6.27), it is possible to show that

$$\|F_j\|_{c_A^0} = \sum_r |\tau_{jr}|. \tag{6.45}$$

Therefore, condition (6.43) is satisfied and, finally, condition (6.22) is fulfilled by Theorem 6.3.

Conversely, we show that if conditions (6.34) and (6.35) are fulfilled and there exists series $\sum_r \tau_{jr}$ satisfying (6.22) such that condition (6.43) holds, then $M \in (c_A, c_B)$. For this, it is sufficient to show that all of the conditions of Theorem 6.3 are satisfied. First we see that the mappings F_j, defined by (6.44) on c_A^0, are continuous linear functionals on c_A^0 by Lemma 5.3, satisfying condition (6.42) by (6.43), and condition (6.36) by (6.45). Thus, $M \in (c_A, c_B)$ by Theorem 6.3.

Example 6.10   Let $A = (a_{nk})$ be a perfect Sq-Sq regular reversible method, $B = (b_{nk})$ is a lower-triangular method and $M = (m_{nk})$ is an arbitrary matrix. We show that then $M \in (c_A, c_B)$ if and only if conditions (6.34) and (6.35) are fulfilled, there exists series $\sum_r \tau_{jr}^n$, satisfying relation (6.31), such that condition (6.32) is satisfied, and

$$\sum_r |D_{nr}| = O(1), \tag{6.46}$$

$$D_{nr} := \sum_{j=0}^{n} b_{nj} \tau_r^j,$$

where series $\sum_r \tau_r^j$, satisfying the property

$$\sum_r |\tau_r^j| = O_j(1), \tag{6.47}$$

is determined by the relation

$$m_{jk} = \sum_r \tau_r^j a_{rk}. \tag{6.48}$$

First, we assume that $M \in (c_A, c_B)$. Then, conditions (6.34) and (6.35) are fulfilled by Theorem 6.2 and, by Proposition 6.3, there exists series $\sum_r \tau_{jr}^n$ satisfying (6.31) such that condition (6.32) is fulfilled.

Further, the functionals f_{jl}, defined by

$$f_{jl}(x) = \sum_{r=0}^{l} \tau_{lr}^j A_r x \tag{6.49}$$

on c_A^0, are continuous and linear on c_A^0 by Lemma 5.3. Besides, it is easy to see that equalities

$$f_{jl}(x) = \sum_{k=0}^{l} m_{jk} x_k \tag{6.50}$$

hold on the fundamental set Δ of c_A^0, and therefore everywhere on c_A^0. This implies that $\lim_l f_{jl}(x) = M_j x$ for each $x \in c_A^0$, and hence the functionals f_j, defined by $f_j(x) := \lim_l f_{jl}(x)$ on c_A^0, belong to $(c_A^0)'$, and $f_j(x) = M_j x$ on c_A^0.

Consequently, by Lemma 5.3, there exists series $\sum_r \tau_r^j$, satisfying condition (6.47), such that

$$f_j(x) = \sum_r \tau_r^j A_r x \qquad (6.51)$$

on c_A^0. For that reason equalities (6.48) hold, and the functionals F_n, defined by (6.37), belong $(c_A^0)'$. Moreover,

$$F_n(x) = \sum_r D_{nr} A_r x \qquad (6.52)$$

for every $x \in c_A^0$. Similarly to the proof of relation (6.27), it is possible to show that

$$\|F_n\|_{c_A^0} = \sum_r |D_{nr}|. \qquad (6.53)$$

Thus, condition (6.46) is fulfilled.

Conversely, assume that conditions (6.34) and (6.35) hold, and there exists series $\sum_r \tau_{jr}^n$, $\sum_r \tau_r^j$, satisfying relations (6.31, 6.32, 6.46–6.48). For $M \in (c_A, c_B)$ it is sufficient to prove that all of the conditions of Theorem 6.2 are fulfilled. First, we see that the transformation $y = Mx$ exists for each $x \in c_A$ by Proposition 6.3. Therefore (see the proof of the necessity), there exists series $\sum_r \tau_r^j$, satisfying (6.47) and (6.52) on c_A^0 with $F_n \in (c_A^0)'$. In addition, the functionals $f_{jl} \in (c_A^0)'$, defined by (6.49), satisfy conditions (6.19) and (6.20) because

$$\|f_{jl}\|_{c_A^0} = \sum_r |\tau_{lr}^j|.$$

The functionals $f_j \in (c_A^0)'$, defined by (6.51), satisfy the relation $f_j(x) = \lim_l f_{jl}(x)$, that is, equalities (6.38) are fulfilled on c_A^0. This implies that $f_j(x) = M_j x$ on c_A^0. As relation (6.53) holds, then it follows from (6.46) that condition (6.36) is satisfied. Thus, all of the conditions of Theorem 6.2 are fulfilled. Consequently, $M \in (c_A, c_B)$.

We note that Theorems 6.2–6.4 were first proved in [2].

6.5 Exercise

Exercise 6.1 To prove that the Cesàro method C^α is perfect for every $\alpha \geq 0$.

Exercise 6.2 Let $A = (a_{nk})$ be an Sr-Sq regular perfect method, such that c_A is a BK-space. Prove that numbers ε_k are convergence factors for c_A if and only if there exist functionals $f_j \in (c_A)'$, such that condition (6.15) is fulfilled and

$$\|f_j\|_{(c_A)'} = O(1).$$

Hint. The proof is similar to the proof of Lemma 6.3.

Exercise 6.3   Let $A = (a_{nk})$ be an Sr-Sq regular reversible perfect method. Show that numbers ε_k are convergence factors for c_A if and only if there exists series $\sum_r \tau_{jr}$, satisfying condition (6.22), and a bounded sequence τ_j such that

$$\tau_j + \sum_r \tau_{jr} a_{rk} = \begin{cases} \varepsilon_k & (k \leq j), \\ 0 & (k > j). \end{cases}$$

Exercise 6.4   Let $A = (a_{nk})$ be an Sr-Sq regular perfect method such that c_A is a BK-space, $B = (b_{nk})$ is a lower-triangular method and $M = (m_{nk})$ is an arbitrary matrix. Prove that $M \in (c_A, c_B)$ if and only if condition (6.35) is satisfied, and there exist functionals $f_{nj} \in (c_A)'$ so that conditions (6.19), (6.21) are satisfied, and

$$\|F_n\|_{(c_A)'} = O(1), \tag{6.54}$$

where the functionals F_n are defined on c_A by equations (6.38) and (6.39).

Hint. The proof is similar to the proof of Theorem 6.2.

Exercise 6.5   Let $A = (a_{nk})$ be an Sr-Sq regular perfect method such that c_A is a BK-space, $B = (b_{nk})$ is a lower-triangular method and $M = (m_{nk})$ is an arbitrary matrix. Prove that A and B are M^{seq}-consistent on c_A if and only if condition (6.34), with $g_k \equiv 0$, is fulfilled, and there exist functionals $f_{nj} \in (c_A)'$, so that conditions (6.19), (6.21), and (6.54) hold.

Hint. The proof is similar to the proof of Theorem 6.4.

Exercise 6.6   Let $A = (a_{nk})$ be an Sq-Sq regular perfect method such that c_A^0 is a BK-AK-space, $B = (b_{nk})$ is a normal method, and $M = (m_{nk})$ is an arbitrary matrix. Prove that A and B are M^{seq}-consistent on c_A if and only if conditions (6.34) and (6.35) hold, with $g_k \equiv 0$ and $g = 1$ are satisfied, and there exist functionals $F_n \in (c_A^0)'$?, so that conditions (6.36) and (6.42) hold.

Hint. Use Theorem 6.3.

Exercise 6.7   Let $A = (a_{nk})$ be a perfect Sr-Sq regular reversible method, $B = (b_{nk})$ is a lower triangular method and $M = (m_{nk})$ is an arbitrary matrix. Prove that $M \in (c_A, c_B)$ if and only if conditions (6.31), (6.33), (6.34), (6.46), $\tau_j^n = O_n(1)$, and

$$\sum_{j=0}^n b_{nj} \tilde{\tau}^j = O(1)$$

are satisfied, where

$$D_{nr} := \sum_{j=0}^n b_{nj} \tilde{\tau}_r^j,$$

and the numbers $\tilde{\tau}^j$ and the sequence $(\tilde{\tau}^j_r) \in l$ (for every fixed j) are determined by the relation

$$m_{jk} = \tilde{\tau}^j + \sum_r \tilde{\tau}^j_r a_{rk}.$$

Hint. Let the mappings F_n be defined by the relation

$$F_n(x) = \left(\sum_{j=0}^n b_{nj} \right) \tau^j \lim_r A_r x + \sum_r D_{nr} A_r x$$

for each $x \in c_A$. If condition (6.46) holds, then $F_n \in (c_A)'$ by Lemma 5.3, and

$$\|F_n\|_{c_A} = \left| \sum_{j=0}^n b_{nj} \tau^j \right| + \sum_r |D_{nr}|.$$

Further, use Exercise 6.4 and Proposition 6.4.

References

1 Aasma, A.: Matrix transformations of summability fields of normal regular matrix methods. Tallinna Tehnikaül. Toimetised. Matem. Füüs. **2**, 3–10 (1994).

2 Aasma, A.: Matrix transformations of summability fields of regular perfect matrix methods. Tartu Ül. Toimetised **970**, 3–12 (1994).

3 Boos, J.: Classical and Modern Methods in Summability. Oxford University Press, Oxford (2000).

4 Jurkat, W.: Summierbarkeitsfaktoren. Math. Z. **58**, 186–203 (1953).

5 Leiger, T.: Funktsionaalanalüüsi meetodid summeeruvusteoorias (Methods of functional analysis in summability theory). Tartu Ülikool, Tartu (1992).

6 Peyerimhoff, A.: Konvergenz- und Summierbarkeitsfaktoren. Math. Z. **55**, 23–54 (1951).

7 Peyerimhoff, A.: Untersuchungen über absolute Summierbarkeit. Math. Z. **57**, 265–290 (1953).

8 Peyerimhoff, A.: Summierbarkeitsfaktoren für absolut Cesàro-summierbare Reihen. Math. Z. **59**, 417–424 (1954).

9 Peyerimhoff, A.: Über Summierbarkeitsfaktoren und verwandte Fragen bei Cesàroverfahren. I. Acad. Serbe Sci. Publ. Inst. Math. **8**, 139–156 (1955).

10 Peyerimhoff, A.: Über Summierbarkeitsfaktoren und verwandte Fragen bei Cesàroverfahren. II. Acad. Serbe Sci. Publ. Inst. Math. **10**, 1–18 (1956).

11 Wilansky, A. Summability through Functional Analysis, North-Holland Mathematics Studies, Vol. **85**; Notas de Matemática (Mathematical Notes), Vol. **91**. North-Holland Publishing Co., Amsterdam (1984).

12 Zeller, K.: Theorie der Limitierungsverfahren. Springer-Verlag, Berlin-Göttingen-Heidelberg (1958).

7

Matrix Transformations of Summability and Absolute Summability Domains: The Case of Special Matrices

7.1 Introduction

All notions and notations not defined in this chapter can be found from Chapters 1, 5, and 6. In this chapter, we continue to study the necessary and sufficient conditions for $M \in (c_A, c_B)$, $M \in (cs_A, cs_B)$, $M \in (bv_A, c_B)$, $M \in (bv_A, bv_B)$, and consider the M^{seq}-consistency of matrix methods A and B on c_A begun in Chapters 5 and 6. We study these problems in cases where A or both A and B are Cesàro (see Section 7.2) or Riesz methods (see Section 7.3). In addition, for some special classes of matrices M, solving the above-mentioned problems, are presented (see Section 7.4).

7.2 The Case of Riesz Methods

In this section, we apply results from Sections 5.3–5.6 and 6.4 for the case if A or both A and B are Riesz methods. Let (p_n) be a sequence of nonzero complex numbers and $P_n = p_0 + \ldots\ldots + p_n \neq 0$, $P_{-1} = 0$. The Riesz method, defined by a lower triangular matrix (a_{nk}), is given in series-to-sequence form by the equations

$$a_{nk} = 1 - P_{k-1}/P_n, \quad k \leq n, \tag{7.1}$$

and, in sequence-to-sequence form, by equalities (cf. with Definition 2.4)

$$a_{nk} = P_k/P_n, \quad k \leq n, \tag{7.2}$$

and, in series-to-series form, by equalities

$$a_{nk} = \frac{P_{k-1} p_n}{P_n P_{n-1}}, \quad k \leq n. \tag{7.3}$$

A Riesz method, defined by (7.1), will be denoted by (R, p_n); a Riesz method, defined by (7.2), will be denoted by $(\widetilde{R}, p_n)$ $((\widetilde{R}, p_n) := (\overline{N}, p_n)$, see Section 2.2);

An Introductory Course in Summability Theory, First Edition. Ants Aasma, Hemen Dutta, and P.N. Natarajan.
© 2017 John Wiley & Sons, Inc. Published 2017 by John Wiley & Sons, Inc.

and a Riesz method, defined by (7.3), will be denoted by $(\widehat{R}, p_n)$. Further, we mean by the phrases "(R, p_n) is a conservative method" and "(R, p_n) is a regular method," correspondingly, that (R, p_n) is Sr-Sq conservative and Sr-Sq regular. Similarly, the phrases "$(\widetilde{R}, p_n)$ is a conservative method" and "$(\widetilde{R}, p_n)$ is a regular method" mean, correspondingly, that $(\widetilde{R}, p_n)$ is Sq-Sq conservative and Sq-Sq regular. By the phrase "(R, p_n) is an absolutely conservative method," we mean that (R, p_n) is Sr-Sq absolutely conservative. The phrases "$(\widehat{R}, p_n)$ is a conservative method" and "$(\widehat{R}, p_n)$ is a regular method" mean, correspondingly, that $(\widehat{R}, p_n)$ is Sr-Sr conservative and Sr-Sr regular. We now present some results, which will be required later.

Lemma 7.1   ([11], p. 114; see also Theorem 2.5) A Riesz method is conservative if and only if $\lim_n P_n$ is not zero and is either finite or infinite, and

$$\sum_{k=0}^{n} |p_k| = O(|P_n|). \tag{7.4}$$

A Riesz method is regular if and only if $\lim_n |P_n| = \infty$ and condition (7.4) is satisfied.

Lemma 7.2   ([11], p. 114) A Riesz method is absolutely conservative if and only if

$$P_{k-1} \sum_{n=k}^{\infty} \left| \frac{p_n}{P_n P_{n-1}} \right| = O(1). \tag{7.5}$$

Theorem 7.1   Let $(\widetilde{R}, p_n)$ be a regular method, $B = (b_{nk})$ a normal method, and $M = (m_{nk})$ an arbitrary matrix. Then, $M \in (c_{(\widetilde{R},p_n)}, c_B)$ if and only if conditions (6.34) and (6.35) are satisfied and

$$g_{jk} = o_j(p_k), \tag{7.6}$$

$$\sum_k \left| P_k \Delta_k \frac{g_{jk}}{p_k} \right| = O(1). \tag{7.7}$$

Proof: **Necessity.** Let $M \in (c_{(\widetilde{R},p_n)}, c_B)$. As $c^0_{(\widetilde{R},p_n)}$ is a BK–AK-space (see [15], p. 39 or [14], p. 117), conditions (6.34) and (6.35) are satisfied due to Example 6.9, and there exists series $\sum_r \tau_{jr}$, satisfying condition (6.22), such that

$$g_{jk} = p_k \sum_{r=k}^{\infty} \frac{\tau_{jr}}{P_r}. \tag{7.8}$$

Hence, condition (7.6) is satisfied, since $\lim_r |P_r| = \infty$ by Lemma 7.1. As

$$\Delta_k \frac{g_{jk}}{p_k} = \Delta_k \left(\sum_{r=k}^{\infty} \frac{\tau_{jr}}{P_r} \right) = \sum_{r=k}^{\infty} \frac{\tau_{jr}}{P_r} - \sum_{r=k+1}^{\infty} \frac{\tau_{jr}}{P_r} = \frac{\tau_{jk}}{P_k},$$

condition (7.7) is also fulfilled.

Sufficiency. Let the conditions of Theorem 7.1 be satisfied. We show that $M \in (c_{(\widetilde{R},p_n)}, c_B)$. Corresponding to Example 6.9, it is sufficient to show that condition (6.43) holds. Let

$$\tau_{jr} := P_k \Delta_k \frac{g_{jk}}{p_k}.$$

Then,

$$\frac{\tau_{jr}}{P_k} = \Delta_k \frac{g_{jk}}{p_k},$$

and condition (6.22) is satisfied by (7.7). Therefore,

$$p_k \sum_{r=k}^{s} \frac{\tau_{jr}}{P_r} = p_k \sum_{r=k}^{s} \Delta_r \frac{g_{jr}}{p_r}$$

$$= p_k \left(\frac{g_{jk}}{p_k} - \frac{g_{j,k+1}}{p_{k+1}} + \frac{g_{j,k+1}}{p_{k+1}} - \frac{g_{j,k+2}}{p_{k+2}} + \cdot + \frac{g_{js}}{p_s} - \frac{g_{j,s+1}}{p_{s+1}} \right)$$

$$= g_{jk} - p_k \frac{g_{j,s+1}}{p_{s+1}}.$$

This implies the validity of (7.8) by condition (7.6). Therefore, condition (6.43) is satisfied, and $M \in (c_{(\widetilde{R},p_n)}, c_B)$.

We shall now consider the (R, p_n) method. It is easy to see that this method is normal and hence has an inverse matrix $A^{-1} = (\eta_{nk})$, which is given by the equations (see [11], p. 116)

$$\eta_{nk} = \begin{cases} P_k/p_k & (n = k), \\ -P_k(1/p_k + 1/p_{k+1}) & (n = k + 1), \\ P_k/p_{k+1} & (n = k + 2), \\ 0 & (n < k \text{ or } n > k + 2). \end{cases} \tag{7.9}$$

Further we need the relationship (proved in [11], p. 58) between $A^{-1} := (\eta_{kl})$ and η_k (defined as in Section 5.2).

Lemma 7.3 If $A = (a_{nk})$ is a normal method, with inverse matrix $A^{-1} := (\eta_{kl})$, such that $a_{n0} \equiv 1$, then $\eta_k = \delta_{k0}$.

Proof: As $A_n e^0 = a_{n0} \equiv 1$, then $Ae^0 = e$. Hence, $e^0 = A^{-1}e$, that is, $\eta_k = \delta_{k0}$.

Using Propositions 5.1 and 5.2, we obtain the following results.

Proposition 7.1 Let (R, p_n) be a conservative method and $M = (m_{nk})$ an arbitrary matrix. Then, the transformation $y = Mx$ exists for every $x \in c_{(R,p_n)}$, if and only if

$$\sum_l \left| P_l \Delta_l \frac{\Delta_l m_{nl}}{p_l} \right| = O_n(1), \tag{7.10}$$

$$P_l m_{nl} = O_n(p_l), \tag{7.11}$$

$$P_l m_{n,l+1} = O_n(p_l). \tag{7.12}$$

Proof: It is sufficient to show that all of the conditions of Proposition 5.1 are fulfilled. Since $a_{n0} \equiv 1$, then $\eta_k = \delta_{k0}$ by Lemma 7.3. Hence, condition (5.15) is satisfied. With the help of (7.9), we obtain

$$
h_{jl}^n = \begin{cases} h_{nl} & (l < j-1), \\ h_{n,j-1} - P_j m_{n,j+1}/p_j + m_{n,j+1} & (l = j-1), \\ P_j m_{nj}/p_j & (l = j), \\ 0 & (l > j), \end{cases} \tag{7.13}
$$

where

$$
h_{nl} = P_l \Delta_l \frac{\Delta_l m_{nl}}{p_l}. \tag{7.14}
$$

Thus, condition (5.14) is fulfilled. As (R, p_n) is conservative, then, by Lemma 7.1, it follows from condition (7.4) that there exists a number $L > 0$, such that $|p_n| < L|P_n|$ for every n. This implies

$$
\left| \frac{P_n}{p_n} \right| > \frac{1}{L}
$$

for all n, that is, the sequence (P_n/p_n) is bounded from below. Consequently, from the validity of (7.11), follows condition

$$
m_{nl} = O_n(1). \tag{7.15}
$$

Now we can conclude that, if condition (5.16) holds, then conditions (7.10)–(7.12) hold by (7.13) and (7.14). Conversely, if conditions (7.10)–(7.12) are fulfilled, then the validity of (5.16) also follows from (7.13) and (7.14). The proof is now complete.

Proposition 7.2 Let (R, p_n) be an absolutely conservative method and $M = (m_{nk})$ an arbitrary matrix. Then, the transformation $y = Mx$ exists for every $x \in bv_{(R,p_n)}$ if and only if conditions (7.11) and (7.12) are fulfilled.

Proof: It is sufficient to show that all of the conditions of Proposition 5.2 are fulfilled. It is easy to see that equalities (7.13) and (7.14) hold. Hence, conditions (5.14) and (5.15) are satisfied (see the proof of Proposition 7.1). Using (7.13) and (7.14), we get

$$
\sum_{l=0}^{r} h_{nl} = \sum_{l=0}^{r} P_l \Delta_l \frac{\Delta_l m_{nl}}{p_l} = \sum_{i=0}^{r} p_i \left(\sum_{l=0}^{r} \Delta_l \frac{\Delta_l m_{nl}}{p_l} \right)
$$

$$
= \sum_{i=0}^{r} \Delta_i m_{ni} - \frac{\Delta_r m_{n,r+1}}{p_{r+1}} \sum_{i=0}^{r} p_i
$$

or

$$
\sum_{l=0}^{r} h_{nl} = m_{n0} - m_{n,r+2} - \frac{P_{r+1}}{p_{r+1}} m_{n,r+1} + \frac{P_{r+1}}{p_{r+1}} m_{n,r+2}, \tag{7.16}
$$

and therefore,

$$\sum_{l=0}^{r} h_{jl}^n = m_{n0} - m_{n,r+2} - \frac{P_{r+1}}{p_{r+1}} m_{n,r+1} + \frac{P_{r+1}}{p_{r+1}} m_{n,r+2}, \text{ if } r < j-1, \tag{7.17}$$

$$\sum_{l=0}^{j-1} h_{jl}^n = m_{n0} - \frac{P_j}{p_j} m_{n,j}, \tag{7.18}$$

$$\sum_{l=0}^{r} h_{jl}^n = m_{n0} \text{ if } r \geq j. \tag{7.19}$$

Moreover, using Lemma 7.2, it is possible to show (similar to the proof of Proposition 7.1) that the sequence (P_n/p_n) is bounded below. Thus, condition (7.15) holds if condition (7.11) is fulfilled. Now we can conclude that, if condition (5.17) holds, then condition (7.11) holds by (7.18); further, condition (7.12) holds by (7.11), (7.15), and (7.17). Conversely, if conditions (7.11) and (7.12) are satisfied, then the validity of (5.17) follows from (7.15) and (7.17)–(7.19). The proof is now complete.

Theorem 7.2 Let (R, p_n) be an absolutely conservative method, $B = (b_{nk})$ a lower triangular method, and $M = (m_{nk})$ an arbitrary matrix. Then, $M \in (bv_{(R,p_n)}, c_B)$ if and only if conditions (6.34), (7.11), and (7.12) are fulfilled and

$$P_r \Delta_r g_{nr} = 0(p_r), \tag{7.20}$$

$$g_{nr} = 0(1). \tag{7.21}$$

Proof: **Necessity.** Let $M \in (bv_{(R,p_n)}, c_B)$. Then, conditions (7.11) and (7.12) are satisfied by Proposition 7.2, and $bv_{(R,p_n)} \subset c_G$. Hence, conditions (6.34) and (7.21) hold by Theorem 1.4, since $l \subset bv_{(R,p_n)}$. With the help of (7.84) (see Exercise 7.1), similar to (7.16), we have

$$\sum_{l=0}^{r} \gamma_{nl} = g_{n0} - g_{n,r+2} - \frac{P_{r+1}}{p_{r+1}} \Delta_r g_{n,r+1}. \tag{7.22}$$

Hence, due to (7.21) and Exercise 5.5, condition (7.20) holds.

Sufficiency. Let all conditions of Theorem 7.2 be satisfied. Then, the transformation $y = Mx$ exists for every $x \in c_{(R,p_n)}$ by Proposition 7.2. This implies that conditions (5.14), (5.15), and (5.17) hold by Proposition 5.2. Moreover, conditions (5.24), (5.25), and (5.45) also are fulfilled. Indeed, condition (5.45) now follows from (7.20) and (7.21). Since $\eta_n = \delta_{n0}$ by Lemma 7.3, conditions (5.24) and (5.25) are satisfied by condition (6.34). Thus, $M \in (bv_{(R,p_n)}, c_B)$ as a result of Exercise 5.5.

Now we consider, as an example, the case if $B = (R, q_n)$ is also the Riesz method, given in the series-to-sequence form, and defined by the numbers q_n.

Example 7.1 Let (R, p_n) and (R, q_n) be conservative methods and $M = (m_{nk})$ an arbitrary matrix. Using Exercises 7.1 and 7.3, it is easy to verify that $M \in (c_{(R,p_n)}, c_{(R,q_n)})$ if and only if conditions (7.11) and (7.12) are fulfilled, and

$$\text{all of the columns of } M \text{ are } (R, q_n)\text{-summable,} \tag{7.23}$$

$$\sum_l \left| P_l \sum_{k=0}^{n} \left(1 - \frac{Q_{k-1}}{Q_n} \right) \Delta_l \frac{\Delta_l m_{kl}}{p_l} \right| = O(1). \tag{7.24}$$

Example 7.2 Let (R, p_n) and (R, q_n) be conservative methods and $M = (m_{nk})$ an arbitrary matrix. With the help of Example 7.1 and Lemma 7.1, we immediately obtain that, if conditions (7.11), (7.12), and (7.23) are satisfied, and

$$\sum_l \sum_k \left| P_l \Delta_l \frac{\Delta_l m_{kl}}{p_l} \right| < \infty, \tag{7.25}$$

then $M \in (c_{(R,p_n)}, c_{(R,q_n)})$.

Example 7.3 Let (R, p_n) and (R, q_n) be absolutely conservative methods and $M = (m_{nk})$ an arbitrary matrix. We show that if conditions (7.11) and (7.12) are fulfilled and

$$P_k \sum_n |\Delta_k m_{nk}| = O(p_k), \tag{7.26}$$

$$\sum_n |m_{nk}| = O(1), \tag{7.27}$$

then $M \in (bv_{(R,p_n)}, c_{(R,q_n)})$. Indeed, as

$$\frac{Q_{k-1}}{Q_n} = O(1) \tag{7.28}$$

by Theorem 1.4 (since $l \subset bv_{(R,q_n)}$), then the finite limits

$$\lim_n \sum_{k=0}^{n} \left| \left(1 - \frac{Q_{k-1}}{Q_n} \right) m_{kl} \right|$$

exist by (7.27). Hence, conditions (6.34) and (7.21) are fulfilled. As

$$\frac{P_k}{p_k} \Delta_k g_{nk} = \frac{P_k}{p_k} \sum_{l=0}^{n} \Delta_k m_{lk} - \frac{P_k}{p_k} \sum_{l=0}^{n} \frac{Q_{l-1}}{Q_n} \Delta_k m_{lk},$$

conditions (7.26) and (7.28) imply the truth of (7.20). Thus, $M \in (bv_{(R,p_n)}, c_{(R,q_n)})$ by Theorem 7.2.

Example 7.4 Let (R, p_n) be an absolutely conservative method, (R, q_n) an arbitrary Riesz method and $M = (m_{nk})$ an arbitrary matrix. Then,

$M \in (bv_{(R,p_n)}, bv_{(R,q_n)})$ if and only if condition (7.11) is fulfilled and

$$\sum_n \left| \frac{q_n}{Q_n Q_{n-1}} \sum_{l=0}^{n} Q_{l-1} m_{lk} \right| = \mathcal{O}(1), \tag{7.29}$$

$$P_k \sum_n \left| \frac{q_n}{Q_n Q_{n-1}} \sum_{l=0}^{n} Q_{l-1} \Delta_k m_{lk} \right| = \mathcal{O}(1), \tag{7.30}$$

where $Q_{-1} = 0$ and $0/0 = 1$. Indeed, this assertion easily follows from Exercise 7.5, since condition (7.12) follows from conditions (7.11) and (7.30).

Now we establish necessary and sufficient conditions for $M \in (cs_{(\hat{R},p_n)}, cs_B)$. For this purpose we need the inverse matrix $A^{-1} = (\eta_{nk})$ of $(\hat{R}, p_n)$, where (see [11], p. 116)

$$\eta_{nk} := \begin{cases} \frac{P_n}{p_n} & (k = n), \\ \frac{P_{n-2}}{p_{n-1}} & (k = n - 1), \\ 0 & (k < n - 1 \text{ or } k > n). \end{cases} \tag{7.31}$$

Theorem 7.3 Let $(\hat{R}, p_n)$ be a conservative method, $B = (b_{nk})$ a lower triangular method, and $M = (m_{nk})$ an arbitrary matrix. Then, $M \in (cs_{(\hat{R},p_n)}, cs_B)$ if and only if conditions (5.66) and (7.11) are satisfied, and

$$P_{l-2} m_{nl} = O_n(p_{l-1}), \tag{7.32}$$

$$\sum_{l=0}^{r} \left| \Delta_l \left(\frac{P_l}{p_l} \Delta_l m_{nl} \right) + \Delta_l m_{n,l+1} \right| = O_n(1), \tag{7.33}$$

$$\sum_l \left| \Delta_l \left(\frac{P_l}{p_l} \sum_{t=0}^{s} \Delta_l g_{nl} \right) + \sum_{n=0}^{r} \Delta_l g_{n,l+1} \right| = O(1). \tag{7.34}$$

Proof: **Necessity.** Assume that $M \in (cs_{(\hat{R},p_n)}, cs_B)$. Then, for $A = (\hat{R}, p_n)$, conditions (5.14), (5.22), (5.47), and (5.48) are satisfied by Theorem 5.5, and condition (5.66) is fulfilled due to Exercise 5.11. With the help of (7.31), we get

$$h_{nl} = \frac{P_l}{p_l} m_{nl} - \frac{P_{l-1}}{p_l} m_{n,l+1}, \tag{7.35}$$

$$h_{jl}^n = \begin{cases} h_{nl} & (l \leq j - 1), \\ \frac{P_j}{p_j} m_{nj} & (l = j), \\ 0 & (l > j). \end{cases}$$

Hence,

$$
\begin{aligned}
\Delta_l h_{jl}^n \big|_{(l \le j-2)} &= h_{jl}^n - h_{j,l+1}^n = \frac{P_l}{p_l} m_{nl} - \frac{P_{l-1}}{p_l} m_{n,l+1} - \frac{P_{l+1}}{p_{l+1}} m_{n,l+1} + \frac{P_l}{p_{l+1}} m_{n,l+2} \\
&= \frac{P_l}{p_l} m_{nl} - \frac{P_{l+1}}{p_{l+1}} m_{n,l+1} - \frac{P_l - p_l}{p_l} m_{n,l+1} + \frac{P_{l+1} - p_{l+1}}{p_{l+1}} m_{n,l+2} \\
&= \frac{P_l}{p_l} \Delta_l m_{nl} - \frac{P_{l+1}}{p_{l+1}} \Delta_l m_{n,l+1} + \Delta_l m_{n,l+1}.
\end{aligned}
$$

This implies that

$$
\Delta_l h_{jl}^n \big|_{(l \le j-2)} = \Delta_l \left(\frac{P_l}{p_l} \Delta_l m_{nl} \right) + \Delta_l m_{n,l+1}. \tag{7.36}
$$

It is easy to see that

$$
\Delta_l h_{j,j-1}^n = \frac{P_{j-1}}{p_{j-1}} m_{n,j-1} - \frac{P_j}{p_j} m_{nj} - \frac{P_{j-2}}{p_{j-1}} m_{nj}, \tag{7.37}
$$

$$
\Delta_l h_{jj}^n = \frac{P_j}{p_j} m_{nj} \tag{7.38}
$$

and

$$
\Delta_l h_{jl}^n \big|_{(l > j)} = 0. \tag{7.39}
$$

Therefore, conditions (7.11) and (7.33) are satisfied as a consequence of (5.22), and

$$
\left| \frac{P_{j-1}}{p_{j-1}} m_{n,j-1} - \frac{P_j}{p_j} m_{nj} - \frac{P_{j-2}}{p_{j-1}} m_{nj} \right| = \mathcal{O}_n(1).
$$

Consequently, condition (7.32) is satisfied by (7.11).

Using (7.35), we get

$$
\gamma_{nl} = \frac{P_l}{p_l} g_{nl} - \frac{P_{l-1}}{p_l} g_{n,l+1}. \tag{7.40}
$$

Therefore, similar to relation (7.36), it is possible to show that

$$
\Delta_l \gamma_{nl} = \Delta_l \left(\frac{P_l}{p_l} \Delta_l g_{nl} \right) + \Delta_l g_{n,l+1}. \tag{7.41}
$$

Thus, condition (7.34) is fulfilled by condition (5.48).

Sufficiency. It is sufficient to show that all of the conditions of Theorem 5.5 are satisfied for $A = (\widehat{R}, p_n)$. First, we see with the help of (5.66) that conditions (5.14) and (5.47) are satisfied by (7.35) and (7.40), respectively. If (7.36)–(7.39) hold, then condition (5.22) is fulfilled by (7.11), (7.32), and (7.33). From relation (7.41), we get, by (7.34), that condition (5.48) is also satisfied. Thus, $M \in (cs_{(\widehat{R}, p_n)}, cs_B)$ by Theorem 5.5. $\qquad \square$

Now we give an application of Theorem 7.3.

Example 7.5 Let $(\widehat{R}, p_n)$ be a regular method, $B = (b_{nk})$ a lower triangular method and $M = (m_{nk})$ an arbitrary matrix. Then, $(\widehat{R}, p_n)$ and B are M^{ser}-consistent on $cs_{(\widehat{R},p_n)}$ if and only if conditions (7.11), (7.32)–(7.34), and (5.66), with $\widehat{g}_k \equiv 1$, are satisfied. Indeed, conditions (5.66), (7.11), and (7.32)–(7.34) are necessary and sufficient for $M \in (cs_{(\widehat{R},p_n)}, cs_B)$ by Theorem 7.3. Hence, conditions (5.14), (5.22), (5.47), and (5.48) are satisfied by Theorem 5.5. By the regularity of $(\widehat{R}, p_n)$, the condition $\widehat{g}_k \equiv 1$ is necessary for the M-consistency of $(\widehat{R}, p_n)$ and B on $cs_{(\widehat{R},p_n)}$. This relation implies, by (7.40), that $\widehat{\gamma}_k \equiv 1$. Consequently, $(\widehat{R}, p_n)$ and B are M^{ser}-consistent on $cs_{(\widehat{R},p_n)}$ by Theorem 5.6.

We note that Theorem 7.2 was first proved in [3] and Theorem 7.3 in [6].

7.3 The Case of Cesàro Methods

In this section, we apply the results from Sections 5.3–5.5 and 6.4 for the case if A or both A and B are Cesàro methods. In this section, we use the notation $A_n^\alpha := A_n^{(\alpha)}$, $\alpha \in \mathbf{C}\backslash\{-1, -2, \dots\}$ (see Section 3.4). The Cesàro method, defined by the lower triangular matrix (a_{nk}), is given (see [11], p. 76, 83–84) in series-to-sequence form by equalities

$$a_{nk} = \frac{A_{n-k}^\alpha}{A_n^\alpha}, \quad k \le n,$$

in sequence-to-sequence form by equalities

$$a_{nk} = \frac{A_{n-k}^{\alpha-1}}{A_n^\alpha}, \quad k \le n, \tag{7.42}$$

and, in series-to-series form, by equalities

$$a_{nk} = \frac{kA_{n-k}^{\alpha-1}}{nA_n^\alpha}, \quad k \le n.$$

We denote a Cesàro series-to-sequence method by C^α, a Cesàro method, given in sequence-to-sequence form, we denote by $\widetilde{C}^\alpha$ (note that, in Section 3.4 it was denoted by (C, α)), and a Cesàro method, given in series-to-series form, was denoted by $\widehat{C}^\alpha$. Further, the phrases "C^α is a conservative method" and "C^α is a regular method" mean, respectively, that C^α is Sr-Sq conservative and Sr-Sq regular. Similarly, the phrases "$\widetilde{C}^\alpha$ is a conservative method" and "$\widetilde{C}^\alpha$ is a regular method" mean, respectively, that $\widetilde{C}^\alpha$ is Sq-Sq conservative and Sq-Sq regular.

The phrases "$\widehat{C}^\alpha$ is a conservative method" and "$\widehat{C}^\alpha$ is a regular method" mean, respectively, that $\widehat{C}^\alpha$ is Sr–Sr conservative and Sr–Sr regular.

It is easy to see that the Cesàro method is normal and therefore has an inverse matrix $A^{-1} = (\eta_{nk})$. For C^α, the lower triangular A^{-1} is given by

$$\eta_{nk} = A_k^\alpha A_{n-k}^{-\alpha-2}, \ k \le n, \tag{7.43}$$

and, for $\widehat{C}^\alpha$, by

$$\eta_{nk} = \frac{k}{n} A_k^\alpha A_{n-k}^{-\alpha-1} \ k \le n. \tag{7.44}$$

(see [11]).

We now present some relations (see [11], pp. 77–79), which will be needed later:

$$A_0^{-1} = 1; \ A_n^{-1} = 0 \text{ for every } n \ge 1, \tag{7.45}$$

$$|A_n^\alpha| \le K_1 (n+1)^{Re\alpha} \text{ for every } \alpha \in \mathbf{C}, K_1 > 0, \tag{7.46}$$

$$|A_n^\alpha| \ge K_2 (n+1)^{Re\alpha} \text{ for } \alpha \in \mathbf{C}\backslash\{-1, -2, \dots\}, K_2 > 0, \tag{7.47}$$

$$\sum_{k=l}^{n} A_{n-k}^\alpha A_{k-l}^\beta = A_{n-l}^{\alpha+\beta+1} \text{ for every } \alpha, \beta \in \mathbf{C}, \tag{7.48}$$

$$\sum_{k=0}^{n} A_{n-k}^\alpha A_k^\beta = A_n^{\alpha+\beta+1} \text{ for every } \alpha, \beta \in \mathbf{C}, \tag{7.49}$$

$$\sum_{k=0}^{n} A_k^\alpha = A_n^{\alpha+1} \text{ for every } \alpha \in \mathbf{C}, \tag{7.50}$$

$$A_n^{\alpha-1} = A_n^\alpha - A_{n-1}^\alpha \text{ for every } \alpha \in \mathbf{C}, \tag{7.51}$$

$$\sum_{k=0}^{n} \frac{A_{n-k}^\sigma}{A_n^\tau} = \frac{\tau}{\tau - \sigma - 1} \frac{1}{A^{\tau-\sigma-1}} \text{ if } Re\tau \ge 0, Re(\tau - \sigma) > 1, \tag{7.52}$$

$$\sum_{k=0}^{n} |A_k^\alpha| = O(1) + O((n+1)^{Re\alpha+1}) \text{ for every } \alpha \in \mathbf{C}, \tag{7.53}$$

$$H^\alpha := \sum_n |A_n^\alpha| < \infty \text{ for } Re\alpha < -1. \tag{7.54}$$

For every bounded sequence (ε_k) and for every $\alpha \in \mathbf{C}$ satisfying relations $Re\alpha > -1$ or $\alpha = -1$, we use the notation

$$\Delta_k^{\alpha+1} \varepsilon_k := \sum_l A_l^{-\alpha-2} \varepsilon_{k+l}.$$

Further, we need the following result.

Lemma 7.4 (see [11], pp. 82–83) The Cesàro method of order α is regular, if $Re\alpha > 0$ or $\alpha = 0$.

The following results deal with the existence of the transform $y = Mx$.

Proposition 7.3 Let α be a complex number with $Re\alpha > 0$ or $\alpha = 0$, and $M = (m_{nk})$ be an arbitrary matrix. Then, the transformation $y = Mx$ exists for every $x \in c_{C^\alpha}$ if and only if

$$\sum_k (k+1)^{Re\alpha} |\Delta_k^{\alpha+1} m_{nk}| = O_n(1),\qquad(7.55)$$

$$m_{nk} = O_n(k^{-Re\alpha}).\qquad(7.56)$$

Proposition 7.4 Let α be a complex number with $Re\alpha > 0$ or $\alpha = 0$, and $M = (m_{nk})$ be an arbitrary matrix. Then, the transformation $y = Mx$ exists for every $x \in c_{\widetilde{C}^\alpha}$ if and only if condition (7.56) is fulfilled and

$$\sum_k (k+1)^{Re\alpha} |\Delta_k^\alpha m_{nk}| = O_n(1).\qquad(7.57)$$

Since C^α and $\widetilde{C}^\alpha$ are regular and $a_{n0} \equiv 1$, then, with help of Lemmas 7.3 and 7.4, the necessity of conditions (7.55)–(7.57) can be easily obtained from Propositions 5.1 and 5.2. The proof of the sufficiency of these conditions is more complicated and, therefore, it will be omitted. The interested reader can find a proof of the sufficiency of (7.55)–(7.57) from [11], pp. 176–183 or from [15], pp. 40–42 (for the existence of $y = Mx$, the numbers m_{nk} for fixed n must be convergence factors for C^α and $\widetilde{C}^\alpha$).

With the help of Example 6.10, we now prove the following result.

Theorem 7.4 Let α be a complex number with $Re\alpha > 0$ or $\alpha = 0$, $B = (b_{nk})$ a lower triangular method and $M = (m_{nk})$ an arbitrary matrix. Then, $M \in (c_{\widetilde{C}^\alpha}, c_B)$ if and only if conditions (7.56), (7.57), (6.34), and (6.35) are fulfilled and

$$\sum_k (k+1)^{Re\alpha} |\Delta_k^\alpha g_{nk}| = O(1).\qquad(7.58)$$

Proof: **Necessity.** Assume that $M \in (c_{\widetilde{C}^\alpha}, c_B)$. Then, conditions (7.56) and (7.57) hold by Proposition 7.4. As $\widetilde{C}^\alpha$ is a perfect method (see, e.g., [16], p. 104 or [12], p. 444), then, by Example 6.10, conditions (6.34) and (6.35) are fulfilled and there exists series $\sum_r \tau_r^j$, satisfying condition (6.47), such that

$$m_{jk} = \sum_{r=k}^\infty \tau_r^j \frac{A_{r-k}^{\alpha-1}}{A_r^\alpha}.$$

Moreover, $A_{r-k}^{\alpha-1}/A_r^\alpha = O(1)$ by (7.46) and (7.47). Hence,

$$\sum_{l=k}^\infty |A_{l-k}^{-\alpha-1}| \sum_{r=l}^\infty \left|\frac{A_{r-l}^{\alpha-1}}{A_r^\alpha}\right| |\tau_r^j| = O_j(1) H^{-\alpha-1} = O_j(1)$$

by (7.54). Therefore, using (7.45) and (7.48), we get

$$\Delta_k^\alpha m_{jk} = \sum_{l=k}^\infty A_{l-k}^{-\alpha-1} \sum_{r=l}^\infty \frac{A_{r-l}^{\alpha-1}}{A_r^\alpha} \tau_r^j = \sum_{r=k}^\infty \frac{\tau_r^j}{A_r^\alpha} \sum_{l=k}^\infty A_{r-l}^{\alpha-1} A_{l-k}^{-\alpha-1}$$

$$= \sum_{r=k}^\infty A_{r-k}^{-1} \frac{\tau_r^j}{A_r^\alpha} = \frac{\tau_k^j}{A_k^\alpha}.$$

This implies that

$$A_k^\alpha \Delta_k^\alpha m_{jk} = \tau_k^j,$$

and, consequently,

$$D_{nk} = A_k^\alpha \Delta_k^\alpha g_{nk}. \tag{7.59}$$

With the help of (7.46), we conclude that condition (7.58) holds due to Example 6.10.

Proof: **Sufficiency.** Let all of the conditions of Theorem 7.4 be satisfied. Then, the transformation $y = Mx$ exists for every $x \in c_{\widetilde{C}^\alpha}$ by Proposition 7.4. Therefore (see Example 6.10), there exists series $\sum_r \tau_r^j$, satisfying condition (6.47), such that relation (6.48) holds. In addition, by Proposition 6.3 there exists series $\sum_r \tau_{jr}^n$, satisfying condition (6.31), such that relation (6.32) holds. Hence (see the proof of the necessity of the present theorem), equalities (7.59) hold. This implies that condition (6.46) is valid by (7.46) and (7.58). Thus, using Example 6.10, we conclude that $M \in (c_{\widetilde{C}^\alpha}, c_B)$. □

Example 7.6 Let α be a real number with $0 \le \alpha \le 1$, $B = (b_{nk})$ a normal method and $M = (m_{nk})$ an arbitrary matrix. Then, $M \in (c_{\widetilde{C}^\alpha}, c_B)$ if and only if conditions (6.34), (6.35), and (7.58) are fulfilled. Indeed $c_{\widetilde{C}^\alpha}^0$ for $0 \le \alpha \le 1$ is a BK-AK space (see [11], pp. 206–210 and Lemma 6.4). Moreover, it is known (see [11], pp. 177–179) that, if α and β are real numbers satisfying the properties $\alpha \ge 0$, $\beta \ge -1$, $\alpha + \beta \ge 0$, and (ε_k) is a sequence of complex numbers satisfying the property $\varepsilon_k = o(1)$, then

$$\Delta_k^{\alpha+\beta} \varepsilon_k = \Delta_k^\beta(\Delta_k^\alpha \varepsilon_k).$$

Hence, using Theorems 6.3 and 7.4, it is possible to prove that $M \in (c_{\widetilde{C}^\alpha}, c_B)$.

For studying the inclusion $M \in (c_{C^\alpha}, c_B)$, we need the following lemma.

Lemma 7.5 Let α be a complex number with $Re\,\alpha > -1$, and let the sequence of complex numbers (ε_k) satisfy the conditions

$$\varepsilon_k = O(1), \tag{7.60}$$

$$\sum_k (k+1)^{Re\alpha} |\Delta_k^{\alpha+1} \varepsilon_k| = O_n(1). \tag{7.61}$$

Then,

$$\lim_k \varepsilon_k = v, \text{ with } v \text{ finite,} \tag{7.62}$$

and

$$\sum_l A_l^\alpha \Delta_k^{\alpha+1} \varepsilon_{k+l} = \varepsilon_k - v. \tag{7.63}$$

Proof: Let a sequence of complex numbers (c_k) be defined by the relation

$$c_k := \sum_l A_l^\alpha \Delta_k^{\alpha+1} \varepsilon_{k+l}.$$

Using (7.46) and (7.47), we conclude that there exists a number $M > 0$, independent of l and k, such that

$$|A_l^\alpha| \leq M|A_{l+k}^\alpha|. \tag{7.64}$$

Consequently,

$$|c_k| \leq M \sum_{l=k}^\infty |A_l^\alpha||\Delta_l^{\alpha+1}\varepsilon_l| \leq MK_1 \sum_{l=k}^\infty (l+1)^{Re\,\alpha}|\Delta_l^{\alpha+1}\varepsilon_l|$$

by (7.46). This implies that

$$\lim_k c_k = 0 \tag{7.65}$$

by (7.61). Moreover, due to (7.51) we get

$$\Delta_k^1 c_k = \Delta_k^{\alpha+1} \varepsilon_k + \sum_{l=k}^\infty (A_l^u - A_{l-1}^u)\Delta_l^{u+1}\varepsilon_{k+l} = \sum_l A_l^{\alpha-1}\Delta_l^{\alpha+1}\varepsilon_{k+l}.$$

In addition, using mathematical induction, it is possible to show that

$$\Delta_k^r c_k = \sum_l A_l^{\alpha-r}\Delta_l^{\alpha+1}\varepsilon_{k+l} \tag{7.66}$$

for every r. Let now $\alpha = s + \sigma + it$, where s and σ are, respectively, the integer and the fractional part of $Re\,\alpha$, and t is the imaginary part of α. Then, from (7.66), for $r = s + 1$ we obtain that the equality

$$\Delta_k^{s+1} c_k = \sum_l A_l^{\sigma+it-1}\Delta_l^{s+\sigma+it+1}\varepsilon_{k+l}$$

is true. If $\sigma = t = 0$ (i.e., α is an integer), then, from the last relation, we get

$$\Delta_k^{s+1} c_k = \Delta_k^{s+1}\varepsilon_k \tag{7.67}$$

by (7.45). Hence (see, e.g., [13], pp. 308–310), there exist numbers $a_0, \ldots, a_s$, such that

$$\varepsilon_k = c_k + a_0 + a_1 k^1 + \ldots + a_s k^s,$$

which implies that $a_1 = \ldots\ldots = a_s = 0$ by (7.60) and (7.65), since $k^s \to \infty$ if $k \to \infty$. Therefore,

$$\varepsilon_k = c_k + a_0. \tag{7.68}$$

We shall now show that (7.67) holds also for $\sigma \neq 0$ or $t \neq 0$. Indeed, by (7.54) and (7.60), we have

$$\sum_j |A_j^{\sigma+it-2}| \sum_l |A_l^{-\alpha-2}||\varepsilon_{k+j+l}| = O(1)H_{\sigma+it-2}H_{-\alpha-2} = O(1),$$

and, consequently, the series

$$\Delta_k^{s+2}c_k = \sum_j \left(\sum_l A_j^{\sigma+it-2}A_l^{-\alpha-2}\varepsilon_{k+j+l} \right)$$

is absolutely convergent. Hence,

$$\Delta_k^{s+2}c_k = \sum_j A_j^{\sigma+it-2} \sum_l A_l^{-\alpha-2}\varepsilon_{k+j+l} = \sum_j A_j^{\sigma+it-2} \sum_{l=j}^{\infty} A_{l-j}^{-\alpha-2}\varepsilon_{k+j}$$

$$= \sum_l \left(\sum_{j=0}^{l} A_j^{\sigma+it-2}A_{l-j}^{-\alpha-2} \right) \varepsilon_{k+l} = \sum_l A_l^{-s-3}\varepsilon_{k+l}$$

by (7.49). Thus, (7.67) (and also (7.68)) holds for $\sigma \neq 0$ or $t \neq 0$. We can conclude that conditions (7.62) and (7.63) are fulfilled for $v = c_0$ by (7.65) and (7.68). $\square$

We note that, for real α, Lemma 7.5 was proved in [10] and, in the general case, in [1].

Theorem 7.5 Let α be a complex number with $Re\alpha > 0$ or $\alpha = 0$, $B = (b_{nk})$ a lower triangular method, and $M = (m_{nk})$ an arbitrary matrix. Then, $M \in (c_{C^\alpha}, c_B)$ if and only if conditions (7.55), (7.56), and (6.34) are fulfilled and

$$\sum_k (k+1)^{Re\alpha}|\Delta_k^{\alpha+1}g_{nk}| = O(1). \tag{7.69}$$

Proof: **Necessity.** Assume that $M \in (c_{C^\alpha}, c_B)$. Then, $c_{C^\alpha} \subset c_G$ (see the proof of Theorem 5.1). Hence, conditions (7.55) and (7.56) are valid by Proposition 7.3, and condition (6.34) is fulfilled by Theorem 1.5, since C^α is regular, by Lemma 7.4. Further, the series $G_n x$ is convergent for every $x \in c_{C^\alpha}$. It remains to show that the numbers g_{nk}, for every fixed n, are the convergence factors for c_{C^α}, from which we have (see [11], p. 192)) that

$$g_{nk} = O_n(k^{-Re\alpha}).$$

Therefore, $g_{nk} = O_n(1)$ (also $\lim_k g_{nk} = 0$). By (7.43), we have

$$\gamma_{nk} = A_k^\alpha \Delta_k^{\alpha+1}g_{nk}. \tag{7.70}$$

Hence, condition (7.69) is valid by Theorem 5.1.

Sufficiency. Assume that all of the conditions of Theorem 7.5 are satisfied and show that, then, all of the conditions of Theorem 5.1 are fulfilled. First, we see that the transformation $y = Mx$ exists for every $x \in c_{C^\alpha}$ by Proposition 7.3, from which we conclude that conditions (5.14)–(5.16) hold. As $A_n^{-j} = 0$ if $n \geq j$ (j a natural number) (see [11], p. 78), then, for a natural number α, condition (5.25) holds because, in this case, the series $\Delta_k^{\alpha+1} g_{nk}$ has only a finite number of summands. We consider now the case where $\alpha \in C \backslash N$ and $Re\ \alpha > 0$. Then, we have

$$\gamma_{r+1,k} - \gamma_{r,k} = A_k^\alpha \Delta_k^{\alpha+1} g_{r+1,k} - A_k^\alpha \Delta_k^{\alpha+1} g_{rk} = A_k^\alpha \sum_i A_i^{-\alpha-2}(g_{r+1,k+i} - g_{r,k+i}).$$

Moreover, as $\lim_k g_{nk} = 0$ (see the proof of the necessity), then, for $\varepsilon_k = g_{nk}$ and $v = 0$, conditions (7.60) and (7.61) are satisfied. This implies that

$$g_{nk} = \sum_l A_l^\alpha \Delta_l^{\alpha+1} g_{n,k+1} = \sum_l \frac{A_l^\alpha}{A_{l+k}^\alpha} \gamma_{n,l+k}$$

by Lemma 7.5. Consequently

$$\sup_{n,k} |g_{nk}| < \infty \tag{7.71}$$

by (7.64) and (7.69). By (7.54), for every $\sigma > 0$ and $k \in N$, there exists a number $L_k^1 > 0$ such that

$$\sum_{i=L_k^1+1}^\infty |A_i^{-\alpha-2}| < \frac{\sigma}{4K_0 |A_k^\alpha|}$$

with $K_0 := \sup_{n,k} |g_{nk}| + 1$. Therefore, the inequality

$$\left| A_k^\alpha \sum_{i=L_k^1+1}^\infty A_i^{-\alpha-2}(g_{r+1,k+i} - g_{r,k+i}) \right| < \frac{\sigma}{2} \tag{7.72}$$

holds independent of l and r. In addition, by (6.34), there exists a number $L_k^2 > 0$ for each fixed k, such that

$$|g_{r+1,k} - g_{r,k}| < \frac{\sigma}{2KH_{-\alpha-2}|A_k^\alpha|}$$

for every $r > L_k^2$ independent of l. The number K is determined by (7.64). Thus, we can conclude that

$$\left| A_k^\alpha \sum_{i=0}^{L_k^1} A_i^{-\alpha-2}(g_{r+1,k+i} - g_{r,k+i}) \right| < \frac{\sigma}{2} \leq |A_k^\alpha| \sum_{i=0}^{L_k^1} |A_i^{-\alpha-2}| |g_{r+1,k+i} - g_{r,k+i}|$$

$$< \frac{\sigma}{2KH_{-\alpha-2}} \sum_{i=0}^{L_k^1} |A_i^{-\alpha-2}| \frac{|A_k^\alpha|}{|A_{k+i}^\alpha|} < \frac{\sigma}{2}$$

for every $r > L_k^2$ independent of l. Hence, by (7.72), we obtain that, for every k the inequalities

$$|g_{r+1,k+i} - g_{r,k+i}| < \sigma$$

hold independent of l if $r > L_k^2$. Thus (γ_{nk}) is a fundamental sequence for every fixed k, and therefore this sequence is convergent, and condition (5.25) is fulfilled.

Also, conditions (5.24) and (5.26) hold. Indeed, (5.26) follows from (7.69). As $\eta_k = \delta_{k0}$ by Lemma 7.3, then the validity of (5.24) follows from the validity of (6.34). Thus, $M \in (c_{C^\alpha}, c_B)$ by Theorem 5.1.

Theorem 7.6 Let α be a complex number with $\mathrm{Re}\,\alpha > 0$ or $\alpha = 0$, $B = (b_{nk})$ a lower triangular method, and $M = (m_{nk})$ an arbitrary matrix. Then, C^α and B are M^{seq}-consistent on c_{C^α} if and only if conditions (7.55), (7.56), (7.69), and (6.34), with $g_k \equiv 1$, are satisfied.

Proof: **Necessity.** Assume that C^α and B are M^{seq}-consistent on c_{C^α}. Then, $M \in (c_{C^\alpha}, c_B)$, and hence conditions (7.55), (7.56), (7.69), and (6.34) are fulfilled by Theorem 7.5. Moreover, C^α and G are consistent on c_{C^α} (see equality (5.33) in the proof of Theorem 5.2). Therefore, by the regularity of C^α (see Lemma 7.4), the method G is also Sr–Sq regular. Thus, $g_k \equiv 1$ by Exercise 1.5.

Sufficiency. We suppose that all of the conditions of the present theorem are fulfilled and show that all conditions of Theorem 5.2 are satisfied. First, we see that $M \in (c_{C^\alpha}, c_B)$ by Theorem 7.5. Hence, conditions (5.14)–(5.16) and (5.24)–(5.26) hold by Theorem 5.2 and $\gamma = g_0 = 1$ since, in the present case, $\eta_n = \delta_{n0}$. Moreover, from (7.52), we get, for $\tau = 0$, that

$$\sum_n A_n^\sigma = 0; \mathrm{Re}\,\sigma < -1,$$

and relation (7.71) holds (see the proof of Theorem 7.5). Therefore, due to (7.54), (7.70), and the condition $g_k \equiv 1$, we obtain

$$\gamma_k = \lim_n \gamma_{nk} = A_k^\alpha \Delta_k^{\alpha+1}(\lim_n g_{nk}) = A_k^\alpha \Delta_k^{\alpha+1} g_k = A_k^\alpha \sum_l A_l^{-\alpha-2} = 0.$$

Consequently, C^α and B are M^{seq}-consistent on c_{C^α} by Theorem 5.2.

We note that, for $B = C^\beta$ (β is a complex number) with $\beta \in \mathbf{C}\backslash\{-1, -2, \dots\}$, from Theorems 7.5 and 7.6, due to Exercise 7.14, the generalization of the results of Alpár from [7–9] follows.

Now we study the inclusions $l_A \subset l_B$ and $l_A \subset cs_B$ in the case, where $A = \widehat{C}^\beta$ and $B \in \mathrm{F}$ (i.e., B is a lower triangular factorable matrix). For this purpose, we use the results of Section 5.5. As $\widehat{C}_n^\alpha e^0 = a_{n0}$ and $a_{n0} = \delta_{n0}$ for

$\alpha \in \mathbf{C}\backslash\{-1, -2, \dots\}$, then $e^0 \in l_{\hat{C}^{\alpha}}$ for $\alpha \in \mathbf{C}\backslash\{-1, -2, \dots\}$. Therefore, from Proposition 5.6, the following are immediately true.

Proposition 7.5 Let α be a complex number with $\alpha \in \mathbf{C}\backslash\{-1, -2, \dots\}$. Then, the following assertions are true:

1. If $l_{\hat{C}^{\alpha}} \subset cs_B$ for $B \in \mathrm{F}$, then $(t_n) \in cs$.
2. If $l_{\hat{C}^{\alpha}} \subset l_M$ for $B \in \mathrm{F}$, then $(t_n) \in l$.

Example 7.7 Let α be a complex number with $Re\alpha > 0$ or $\alpha = 0$, and $u = (u_k)$ defined by $u_k = 1/A_k^r$, $r \in C$. We show that $l_{\hat{C}^{\alpha}} \subset cs_B$ for each $B \in \mathrm{F}_u^{cs}$ if and only if $Re\alpha \le Re r$. Indeed, condition (5.60) can be presented by the relation

$$V_l := lA_l^{\alpha} \sum_{n=l}^{\infty} \left| \frac{A_{n-l}^{-\alpha-1}}{nA_n^r} \right| = O(1). \tag{7.73}$$

Thus, by Theorem 5.8, $l_{\hat{C}^{\alpha}} \subset cs_B$ for each $B \in \mathrm{F}_u^{cs}$ if and only if condition (7.73) holds. For $\alpha = 0$, this condition is equivalent to

$$\frac{1}{|A_l^r|} = O(1) \tag{7.74}$$

by (7.45) since $A_l^0 = 1$. Condition (7.74) holds, due to (7.46) and (7.47), if and only if $Re r \ge 0$. So the proof is completed for $\alpha = 0$.

Let further $Re\alpha > 0$. By Corollary 5.4, the validity of condition (5.62), which can be presented in the form

$$\left| \frac{A_l^{\alpha}}{A_l^r} \right| = O(1) \tag{7.75}$$

by (7.45), is necessary for $l_{\hat{C}^{\alpha}} \subset cs_B$ for each $B \in \mathrm{F}_u^{cs}$. Condition (7.75) holds if and only if $Re\alpha \le Re r$ by (7.46) and (7.47). Hence, with the help of (7.46) and (7.47), we obtain for $Re\alpha \le Re t$ that

$$V_l = O(1)(l+1)^{Re\alpha+1} \sum_{n=l}^{\infty} \frac{(n-l+1)^{-Re\alpha-1}}{(n+1)^{Re r+1}}$$

$$= O(1)(l+1)^{Re\alpha+1} \sum_{n=0}^{\infty} \frac{1}{(n+1)^{Re\alpha+1}(n+l+1)^{Re r+1}}$$

$$= O(1)(l+1)^{Re(\alpha-r)} \sum_{n=0}^{\infty} \frac{1}{(n+1)^{Re\alpha+1}(\frac{n}{l+1}+1)^{Re r+1}}$$

$$= O(1) \sum_{n=0}^{\infty} \frac{1}{(n+1)^{Re\alpha+1}} = O(1),$$

that is, condition (7.73) holds. The proof is complete.

We note that Theorem 7.5 was first proved in [1] and Example 7.7 in [5].

7.4 Some Classes of Matrix Transforms

In this section, we describe some classes of matrices M, belonging to (c_A, c_B), for given matrix methods A and B.

First, for the case of lower triangular matrices $A = (a_{nk})$, we consider the following classes of matrices:

$$S_A := \left\{ M = (m_{nk}) : m_{nk} = \sum_{l=k}^{n} s_{nl} a_{lk}, S = (s_{nk}) \text{ is lower triangular} \right\},$$

$$V_A := \{ M = (m_{nk}) : m_{nk} = v_n a_{nk}, v_n \in \mathbf{C} \}.$$

Theorem 7.7 Let A be a triangular matrix and B an Sq-Sq conservative matrix. Then, $M = SA \in S_A$ belongs (c_A, c_B) for every lower triangular matrix $S = (s_{nk})$ satisfying conditions

$$s_k := \lim_n s_{nk} \text{ exists and is finite,} \tag{7.76}$$

$$s := \lim_n \sum_{k=0}^{n} s_{nk} \text{ exists and is finite,} \tag{7.77}$$

$$\sum_{k=0}^{n} |s_{nk}| = O(1). \tag{7.78}$$

Proof: It is easy to see that

$$M_n x = \sum_{v=0}^{n} s_{nv} A_v x$$

for every $x \in c_A$. In addition, if S satisfies conditions (7.76)–(7.78), then S is Sq-Sq conservative by Exercise 1.1. This implies that $(M_n x) \in c$ for every Sq-Sq conservative S since $(A_v x) \in c$ for each $x \in c_A$. Therefore, $(M_n x) \in c_B$ for every $x \in c_A$ because B is Sq-Sq conservative. $\qquad \square$

Example 7.8 Let A be an Sq-Sq regular or an Sr–Sq regular triangular method, B an Sq-Sq regular method, and $M = SA \in S_A$. Then, from Theorem 7.7, it follows by Theorem 1.1 that A and B are M^{seq}-consistent on c_A, if $S = (s_{nk})$ satisfies conditions (7.76)–(7.78) with $s_k = 0$ and $s = 1$.

Example 7.9 Let A be a triangular method, B an Sq–Sq conservative method, and $M = (v_n a_{nk}) \in V_A$. Then, it is easy to see that $M \in (c_A, c_B)$ for every $(v_n) \in c$.

Example 7.10 Let A be a normal method, B a triangular method, and $M = SA \in S_A$, where $S = (s_{nk})$ is a lower triangular matrix. With the help of Exercise 1.1, it is easy to show that $M \in (c_A, c_B)$ if and only if the matrix $T = (t_{nk})$, defined by

$$t_{nk} := \sum_{l=k}^{n} b_{nl} s_{lk},$$

satisfies conditions

$$t_k := \lim_n t_{nk} \text{ exists and is finite}, \tag{7.79}$$

$$t := \lim_n \sum_{k=0}^{n} t_{nk} \text{ exists and is finite}, \tag{7.80}$$

$$\sum_{k=0}^{n} |t_{nk}| = O(1). \tag{7.81}$$

Example 7.11 Let all of the assumptions of Example 7.10 be satisfied. Using Theorem 1.1, it is easy to show that A and B are M^{seq}-consistent on c_A if T satisfies conditions (7.79)–(7.81) with $t_k = 0$ and $t = 1$.

Example 7.12 Let A be a normal method, β a complex number with $Re\beta > 0$ or $\beta = 0$ and $M^\tau = S^\tau A \in S_A$ where $S^\tau = (s_{nk})$ is defined by a lower triangular matrix $s_{nk} := A_{n-k}^\tau$ (τ is a complex number). Using Example 7.10, we prove that $M^\tau \in (c_A, c_{\tilde{C}^\beta})$ if $\tau \leq -1$. We will show that conditions (7.79)–(7.81) hold for $B = \tilde{C}^\beta$ and $M = M^\tau$. By (7.48) and (7.50), we obtain

$$t_{nk} = \frac{1}{A_n^\beta} \sum_{l=k}^{n} A_{n-l}^{\beta-1} A_{l-k}^\tau = \frac{A_{n-k}^{\beta+\tau}}{A_n^\beta}$$

and

$$\sum_{k=0}^{n} t_{nk} = \frac{1}{A_n^\beta} \sum_{k=0}^{n} A_{n-k}^{\beta+\tau} = \frac{A_n^{\beta+\tau+1}}{A_n^\beta}.$$

We get

$$\left| \frac{A_{n-k}^{\beta+\tau}}{A_n^\beta} \right| = O(1) \frac{(n-k+1)^{Re\beta+\tau}}{(n+1)^{Re\beta}} = O(1)\left(1 - \frac{k}{n+1}\right)^{Re\beta}(n-k+1)^\tau = o(1)$$

by (7.46) and (7.47). This implies that $\lim_n t_{nk} = 0$ for every k, which means that condition (7.79) is satisfied.
For $\tau < -1$,

$$\left| \frac{A_n^{\beta+\tau+1}}{A_n^\beta} \right| = O(1)(n+1))^{\tau+1} = o(1)$$

by (7.46) and (7.47). If $\tau = -1$, then

$$\frac{A_n^{\beta+\tau+1}}{A_n^\beta} = 1.$$

Hence,

$$\lim_n \sum_{k=0}^n t_{nk} = 0, \tau < -1, \text{ and } \lim_n \sum_{k=0}^n t_{nk} = 1, \tau = -1,$$

that is, condition (7.80) holds.

Condition (7.81) is also satisfied. Indeed, let at first $Re\beta > 0$. If, in addition, $Re\beta + \tau \neq -1$, then using (7.47) and (7.53), we conclude that

$$\sum_{k=0}^n |t_{nk}| = \frac{1}{|A_n^\beta|} \sum_{k=0}^n |A_{n-k}^{\beta+\tau}| = \frac{1}{|A_n^\beta|} \sum_{k=0}^n |A_k^{\beta+\tau}|$$

$$= \frac{O(1) + O((n+1)^{Re\beta+\tau+1})}{(n+1)^{Re\beta}} = O(1).$$

For $Re\beta + \tau = -1$, using (7.46) and (7.47), we get

$$\sum_{k=0}^n |t_{nk}| = O(1)\frac{1}{(n+1)^{Re\beta}} \sum_{k=0}^n (n-k+1)^{-1} = O(1)\frac{\ln(n+1)}{(n+1)^{Re\beta}} = o(1).$$

Consequently, condition (7.81) holds for $Re\beta > 0$.

Let now $\beta = 0$. If, in addition, $\tau < -1$, then, due to the equality $A_n^0 = 1$, we obtain

$$\sum_{k=0}^n |t_{nk}| = \sum_{k=0}^n |A_{n-k}^\tau| = O(1)$$

by (7.54). For $\tau = -1$ we have

$$\sum_{k=0}^n |t_{nk}| = 1$$

by (7.45). Hence, condition (7.81) is also satisfied for $Re\beta > 0$.

Finally, we state one result for the special case where $M = (m_{nk})$ is a lower triangular matrix with

$$m_{nk} = A_{n-k}^r; \ r \in \mathbf{C}. \tag{7.82}$$

Theorem 7.8 Let $\alpha, \beta \in \mathbf{C}\backslash\{-1, -2, \dots\}$ and r be a complex number. If $Rer < -1$ and $Rer < Re\alpha \leq Re\beta$, then $M = (m_{nk})$, defined by (7.82), transforms c_{C^α} into c_{C^β}.

We note that the proof of this result, provided in [4], is long and troublesome; therefore, we will omit it.

7.5 Exercise

Exercise 7.1   Let (R, p_n) be a conservative method, $B = (b_{nk})$ a lower triangular method, and $M = (m_{nk})$ an arbitrary matrix. Prove that $M \in (c_{(R,p_n)}, c_B)$ if and only if conditions (7.10)–(7.12), and (6.34) are fulfilled and

$$\sum_l \left| P_l \Delta_l \frac{\Delta_l g_{nl}}{p_l} \right| = O(1). \tag{7.83}$$

Hint. With the help of (7.9), we obtain

$$\gamma_{nl} = P_l \Delta_l \frac{\Delta_l g_{nl}}{p_l}. \tag{7.84}$$

Further, use Proposition 7.1, Theorems 1.5 and 5.1, and Lemma 7.3.

Exercise 7.2   Let (R, p_n) be a regular method, $B = (b_{nk})$ a lower triangular method and $M = (m_{nk})$ an arbitrary matrix. Prove that (R, p_n) and B are M^{seq}-consistent on $c_{(R,p_n)}$ if and only if conditions (7.10)–(7.12), (7.83), and (6.34) with $g_k \equiv 0$ are fulfilled.

Hint. Use Theorem 5.2 for $A = (R, p_n))$ and Theorem 1.5.

Exercise 7.3   Prove that if B is a normal method, then condition (7.10) is redundant in Exercises 7.1 and 7.2.

Hint. That (7.10) is true follows from the truth of (7.83). For $n = 0$, (7.83) implies

$$\sum_l \left| P_l \Delta_l \frac{\Delta_l m_{0l}}{p_l} \right| < \infty,$$

and for $n = i$,

$$T_i := \sum_l |P_l| \left| b_{i0} \Delta_l \frac{\Delta_l m_{0l}}{p_l} + \cdot + b_{ii} \Delta_l \frac{\Delta_l m_{il}}{p_l} \right| < \infty.$$

Since B is normal, we can conclude that

$$\sum_l |P_l| \left| \Delta_l \frac{\Delta_l m_{il}}{p_l} \right| \leq \frac{1}{b_{ii}} \left(T_i + \sum_{k=0}^{i-1} |b_{ik}| \sum_l |P_l| \left| \Delta_l \frac{\Delta_l m_{kl}}{p_l} \right| \right).$$

Then, use mathematical induction.

Exercise 7.4   Let (R, p_n) be a conservative method, $B = (b_{nk})$ a lower triangular method, and $M = (m_{nk})$ an arbitrary matrix. Prove that $M \in (c_{(R,p_n)}, bv_B)$ if and only if conditions (7.10)–(7.12) are fulfilled and

$$\sum_n |g_{n0} - g_{n-1,0}| < \infty, \tag{7.85}$$

$$\sum_{n \in K} \sum_{k \in L} P_k \Delta_k \frac{\Delta_l(g_{nk} - g_{n-1,k})}{p_k} = O(1), \tag{7.86}$$

where $g_{-1,k} \equiv 0$, K, and L are arbitrary finite subsets of **N**.

Hint. Use Exercise 5.7.

Exercise 7.5   Let (R, p_n) be an absolutely conservative method, $B = (b_{nk})$ a lower triangular method, and $M = (m_{nk})$ an arbitrary matrix. Prove that $M \in (bv_{(R,p_n)}, bv_B)$ if and only if conditions (7.11) and (7.12) are fulfilled and

$$\sum_n |g_{nk} - g_{n-1,k}| = O(1), \tag{7.87}$$

$$P_k \sum_n |\Delta_k(g_{nk} - g_{n-1,k})| = O(p_k), \tag{7.88}$$

where $g_{-1,k} \equiv 0$.

Hint. Use Exercises 1.11 and 5.6.

Exercise 7.6   Let (R, p_n) be a conservative method, $B = (b_{nk})$ a method satisfying condition (5.36), and $M = (m_{nk})$ an arbitrary matrix. Prove that if conditions (6.34) and (7.83) are fulfilled and

$$\sum_l \left| P_l \Delta_l \frac{\Delta_l m_{nl}}{p_l} \right| = O(1), \tag{7.89}$$

$$P_l m_{nl} = O(p_l), \tag{7.90}$$

$$P_l m_{n,l+1} = O(p_l), \tag{7.91}$$

then $M \in (c_{(R,p_n)}, c_B)$.

Hint. Show that conditions (5.36), (7.83), and (7.89)–(7.91) imply the relation

$$m_{nl} = O(l). \tag{7.92}$$

Then, prove that all of the assumptions and conditions of Theorem 5.3 are satisfied.

Exercise 7.7   Let (R, p_n) be a conservative method, $B = (b_{nk})$ a method satisfying condition (5.36), and $M = (m_{nk})$ an arbitrary matrix. Prove that if conditions (7.85), (7.86), and (7.89)–(7.91) are satisfied, then $M \in (c_{(R,p_n)}, bv_B)$.

Hint. Use Exercise 5.10.

Exercise 7.8 Let (R, p_n) be an absolutely conservative method, $B = (b_{nk})$ a method satisfying condition (5.36), and $M = (m_{nk})$ an arbitrary matrix. Prove that if conditions (6.34), (7.20), (7.21), (7.90), and (7.91) are satisfied, then $M \in (bv_{(R,p_n)}, c_B)$.

Hint. Use Theorem 5.4.

Exercise 7.9 Let (R, p_n) be an absolutely conservative method, $B = (b_{nk})$ a method, satisfying condition (5.36), and $M = (m_{nk})$ an arbitrary matrix. Prove that if conditions (7.87), (7.88), (7.90), and (7.91) are satisfied, then $M \in (bv_{(R,p_n)}, bv_B)$.

Hint. Use Exercise 5.9.

Exercise 7.10 Prove that if a method B has the property $\sum_k |b_{nk}| = O(1)$, then condition (7.83) is redundant in Exercise 7.6, and conditions (7.20) and (7.21) in Exercise are true.

Hint. See Hint of Exercise 7.3.

Exercise 7.11 Let $(\widehat{R}, p_n)$ be a conservative method, $B = (b_{nk})$ a lower triangular method and $M = (m_{nk})$ an arbitrary matrix. Prove that if $M \in (cs_{(\widehat{R},p_n)}, cs_B)$, then

$$m_{nl} = O_n(p_l) \text{ and } m_{nl} = O_n(p_{l-1}).$$

Hint. Use Theorem 7.3 and Lemma 7.1.

Exercise 7.12 Let $(\widehat{R}, p_n)$ be a regular method, $B = (b_{nk})$ a lower triangular method and $M = (m_{nk})$ an arbitrary matrix. Prove that if $M \in (cs_{(\widehat{R},p_n)}, cs_B)$, then

$$m_{nl} = o_n(p_l) \text{ and } m_{nl} = o_n(p_{l-1}).$$

Exercise 7.13 Show that the method $\widetilde{C}^\alpha$, defined by (7.42), coincides with the method (C, α), defined by the matrix (c_{nk}^α) in Section 3.4.

Exercise 7.14 Prove that for a normal method B, conditions (7.55) and (7.57) are redundant in Theorems 7.4–7.6.

Hint. See Hint of Exercise 7.3.

Exercise 7.15 Let α be a complex number with $Re\alpha > 0$ or $\alpha = 0$, and $u = (u_k)$ be defined by $u_k = 1/A_k^r$, $r \in \mathbf{C}$. Prove that $l_{\widetilde{C}^\alpha} \subset l_B$ for each $B \in F_u^l$ if and only if $Re\alpha \leq Rer$.

Hint. Use Example 7.7 and Corollary 5.3.

Exercise 7.16   Let α be a complex number with $Re\alpha > 0$ or $\alpha = 0$, and $B = (b_{nk})$ be a method satisfying the property $l \subset c_B$. Let (u_k) be defined by $u_k := 1/A_k^r$, where r is a complex number satisfying the conditions $Rer > 0$, and $(t_n) \in l$. Prove that $M = (t_n u_k) \in$ F belongs to (c_{C^α}, c_B) if $Re\alpha \leq Rer$.

Hint. Prove, with the help of Example 5.5, that condition (5.54) holds for $A = C^\alpha$ and $u_k = 1/A_k^r$. Use it to establish relations (7.47) and (7.52). Then, apply Example 5.6.

Exercise 7.17   Let α be a complex number with $Re\alpha > 0$ or $\alpha = 0$, and $B = (b_{nk})$ be a method satisfying the property $l \subset c_B$. Let (u_k) be defined by $u_k := d^k$, where d is a real number, and $(r_n) \in l$. Prove that $M = (t_n u_k) \in$ F belongs (c_{C^α}, c_B) if $|d| < 1$.

Exercise 7.18   Let A be an Sq-Sq regular or an Sr–Sq regular triangular method, B an Sq-Sq regular matrix, and $M = (v_n a_{nk}) \in V_A$. Prove that A and B are M^{seq}-consistent on c_A if $\lim_n v_n = 1$.

Exercise 7.19   Let A be a normal method, B an arbitrary method, and $M = (v_n a_{nk}) \in V_A$. Prove that $M \in (c_A, c_B)$ if and only if the matrix $\overline{B} = (\overline{b}_{nk})$, where

$$\overline{b}_{nk} := b_{nk} v_k,$$

is Sq-Sq conservative.

Exercise 7.20   Let all of the assumptions of Exercise 7.19 be satisfied. Prove that if $\overline{B}$ is Sq–Sq regular, then A and B are M^{seq}-consistent on c_A.

Exercise 7.21   Prove that the matrix $\overline{B}$ is Sq–Sq regular in Exercise 7.20, if B is Sq–Sq regular and $\lim_n v_n = 1$.

References

1 Aasma, A.: Preobrazovanija polei summirujemosti (Matrix transformations of summability fields). Tartu Riikl. Ül. Toimetised **770**, 38–50 (1987).
2 Aasma, A.: Characterization of matrix transformations of summability fields. Tartu Ül. Toimetised **928**, 3–14 (1991).
3 Aasma, A.: On the matrix transformations of absolute summability fields of reversible matrices. Acta Math. Hung. **64(2)**, 143–150 (1994).
4 Aasma, A.: Some notes on matrix transforms of summability domains of Cesàro matrices. Math. Model. Anal. **15(2)**, 153–160 (2010).

5 Aasma, A.: Some inclusion theorems for absolute summability. Appl. Math. Lett. **25**(3), 404–407 (2012).

6 Aasma, A.: Matrix transforms of summability domains of normal series-to-series matrices. J. Adv. Appl. Comput. Math. **1**, 35–39 (2014).

7 Alpár, L.: Sur certains changements de variable des séries de Faber (Certain changes of variables in Faber series). Stud. Sci. Math. Hung. **13**(1-2), 173–180 (1978).

8 Alpár, L.: Cesàro Summability and Conformal Mapping, Functions, Series, Operators, Vols **I, II**. Budapest, 1980, 101–125; Colloq. Math. Soc. János Bolyai, Vol. **35**. North-Holland, Amsterdam (1983).

9 Alpár, L.: On the linear transformations of series summable in the sense of Cesàro. Acta Math. Hung. **39**(1-2), 233–243 (1982).

10 Andersen, A.F.: On the extensions within the theory of Cesàro summability of a classical convergence theorem of Dedekind. Proc. London Math. Soc. **8**, 1–52 (1958).

11 Baron, S.: Vvedenie v teoriyu summiruemosti ryadov (Introduction to the Theory of Summability of Series). Valgus, Tallinn (1977).

12 Boos, J.: Classical and Modern Methods in Summability. Oxford University Press, Oxford (2000).

13 Gelfond, A.: Istsislenie konecnŏh raznostei (Computation of finite differences). Nauka, Moscow (1967).

14 Kangro, G.: λ-perfektnost metodov summirovanya i priloženya. I. (The λ-perfectness of summability methods and applications of it. I.). Eesti NSV Tead. Akad. Toimetised Füüs.-Mat. **20**, 111–120 (1971).

15 Peyerimhoff, A.: Konvergenz- und Summierbarkeitsfaktoren. Math. Z. **55**, 23–54 (1951).

16 Zeller, K.: Theorie der Limitierungsverfahren. Springer-Verlag, Berlin-Göttingen-Heidelberg (1958).

8

On Convergence and Summability with Speed I

8.1 Introduction

In studying a convergent process, it is often important to estimate the speed of convergence of this process. For example, in the theory of approximation, and using numerical methods for solving differential and integral equations, several methods have been worked out for estimating the speed of convergence. The interested reader can find a rather overwhelming overview, for example, from Refs. [10, 11, 13], and [28]. In this chapter, we describe the method introduced by Kangro (see [17, 19]). Before formulating the basic notions, we note that all of the notions not defined in this chapter can be found in Chapters 5–7. Also throughout this chapter, we suppose that $\lambda = (\lambda_k)$ is a sequence with $0 < \lambda_k \nearrow \infty$, if not specified otherwise.

Definition 8.1   A convergent sequence $x := (\xi_k)$ with

$$\lim_k \xi_k := \varsigma \text{ and } v_k := \lambda_k(\xi_k - \varsigma) \tag{8.1}$$

is called bounded with the speed λ, for brevity, λ-bounded if $v_k = O(1)$, and convergent with speed λ, for brevity, λ-convergent if the limit $\lim_k v_k$ exists and is finite.

The set of all λ-bounded sequences is denoted by m^λ, and the set of all λ-convergent sequences by c^λ. It is easy to see that $c^\lambda \subset m^\lambda \subset c$, and, if $\lambda_k = O(1)$, then $c^\lambda = m^\lambda = c$. Let $A = (a_{nk})$ be a matrix method over the real or complex numbers, and $\mu = (\mu_k)$ another speed; that is, $0 < \mu_k \nearrow \infty$. In this chapter, we present necessary and sufficient conditions for $A \in (m^\lambda, m^\mu)$, $A \in (c^\lambda, c^\mu)$ and $A \in (c^\lambda, m^\mu)$, were first established by Kangro [17–19].

Definition 8.2   A sequence $x = (x_k)$ is said to be A^λ-bounded (A^λ-summable), if $Ax \in m^\lambda$ ($Ax \in c^\lambda$, respectively).

An Introductory Course in Summability Theory, First Edition. Ants Aasma, Hemen Dutta, and P.N. Natarajan.
© 2017 John Wiley & Sons, Inc. Published 2017 by John Wiley & Sons, Inc.

The set of all A^λ-bounded sequences is denoted by m_A^λ, and the set of all A^λ-summable sequences by c_A^λ. Of course, $c_A^\lambda \subset m_A^\lambda \subset c_A$, and, if λ is a bounded sequence, then $m_A^\lambda = c_A^\lambda = c_A$. We also note that some modifications of Kangro's definitions (given by Definitions 5.1 and 5.2) were made by Sikk (see [29–33]) and Jürimäe (see [15, 16]). We do not consider these modifications in this chapter.

Let $B = (b_{nk})$ and $M = (m_{nk})$ be matrices with real or complex entries.

Definition 8.3 Matrices A and B are said to be M^{seq}-consistent on m_A^λ, or on c_A, if the transformation Mx exists and

$$\lim_n B_n(Mx) = \lim_n A_n x$$

for each $x \in m_A^\lambda$ or $x \in c_A$, respectively.

In this chapter, we also introduce necessary and sufficient conditions for $M \in (m_A^\lambda, m_B^\mu)$ and for the M^{seq}-consistency of A and B on m_A^λ. Necessary and sufficient conditions for $M \in (m_A^\lambda, m_B^\mu)$ and for the M^{seq}-consistency of A and B on m_A^λ, if A is a normal method and B is a triangular method were found by Aasma [2]. If a matrix has the form (5.1), where (ε_k) is a sequence of numbers, then the above-mentioned problem reduces to the problem of finding necessary and sufficient conditions for numbers ε_k to be the B^μ-boundedness factors for m_A^λ. We say that numbers ε_k are the B^μ-boundedness factors for m_A^λ if $(\varepsilon_k x_k) \in m_B^\mu$ for every $x := (x_k) \in m_A^\lambda$. If B is the identity method, that is, $B = E$, and $\mu_k = O(1)$, then we obtain necessary and sufficient conditions for numbers ε_k to be the convergence factors for m_A^λ. Necessary and sufficient conditions for numbers ε_k to be the B^μ-boundedness factors for m_A^λ were obtained by Kangro [19]. If, in addition, we take $\varepsilon_k \equiv 1$, then $m_{nk} = \delta_{nk}$; that is, $Mx = x$ for each $x \in m_A^\lambda$. So we get the inclusion problem $m_A^\lambda \subset m_B^\mu$ studied by Leiger and Maasik (see [25]).

The B^μ-boundedness factors and the convergence factors for m_A^λ have been used for estimating the rate of convergence of a Fourier series and general orthogonal series (see, e.g., [5–9, 20–22]). However, Aljanăić [5–9] did not use directly the notions of convergence and summability with speed, which have been introduced later by Kangro. Also Sikk [29–31] used his modified concepts of the convergence and summability with the speed to estimate the rate of convergence of Fourier series. The results on $M \in (m_A^\lambda, m_B^\mu)$ were used by Aasma for the comparison of the order of approximation of Fourier expansions in Banach spaces by different matrix methods (see [1–3]).

Later Aasma complemented the Kangro's concepts of the boundedness with speed and the A-boundedness with speed. He obtained a new method for the estimating, accelerating and comparing the speeds of convergence of series and sequences. In this chapter, we do not consider this topic (nor the convergence acceleration problems). The interested reader may refer [4]. We also note that the estimation and the comparison of speeds of convergence of series and sequences, based on Kangro's concepts of convergence, boundedness, and summability with speed, have also been studied by Šeletski and Tali [26, 27], Stadtmüller and Tali [34], and Tammeraid (see [35–38]). Tammeraid generalized the concepts of A^λ-summability and A^λ-boundedness, considering a matrix whose elements are bounded linear operators from a Banach space X into a Banach space Y.

8.2 The Sets (m^λ, m^μ), (c^λ, c^μ), and (c^λ, m^μ)

In this section, we present necessary and sufficient conditions for $A \in (m^\lambda, m^\mu)$, $A \in (c^\lambda, c^\mu)$, and $A \in (c^\lambda, m^\mu)$, if $A = (a_{nk})$ is an arbitrary matrix with real or complex entries, and $\lambda := (\lambda_k), \mu := (\mu_k)$ are monotonically increasing sequences with $\lambda_k > 0$ and $\mu_k > 0$.

Theorem 8.1 A method $A = (a_{nk}) \in (m^\lambda, m^\mu)$ if and only if condition (1.2) is satisfied and

$$Ae \in m^\mu, \tag{8.2}$$

$$\sum_k \frac{|a_{nk}|}{\lambda_k} = O(1), \tag{8.3}$$

$$\mu_n \sum_k \frac{|a_{nk} - \delta_k|}{\lambda_k} = O(1). \tag{8.4}$$

If also $\mu_n = O(1)$ and $\lambda_n \neq O(1)$, then, in (8.4), it is necessary to replace $O(1)$ by $o(1)$.

Proof: **Necessity.** Assume that $A \in (m^\lambda, m^\mu)$. It is easy to see that $e^k \in m^\lambda$ and $e \in m^\lambda$. Hence, conditions (1.2) and (8.2) hold. Since, from (8.1) we have

$$\xi_k = \frac{v_k}{\lambda_k} + \varsigma; \; \varsigma := \lim_k \xi_k, (v_k) \in m$$

for every $x := (\xi_k) \in m^\lambda$, it follows that

$$A_n x = \sum_k \frac{a_{nk}}{\lambda_k} v_k + \varsigma \mathfrak{A}_n; \; \mathfrak{A}_n := \sum_k a_{nk}. \tag{8.5}$$

As $(\mathfrak{A}_n) \in m^\mu$ by (8.2), then, from (8.4) we can conclude that the method

$$A_\lambda := \left(\frac{a_{nk}}{\lambda_k} \right)$$

transforms this bounded sequence (v_k) into c.

Assume now that $\lambda_n \neq O(1)$. Then, for every sequence $(v_k) \in m$, the sequence $(v_k/\lambda_k) \in c_0$. But, for (v_k/λ_k), there exists a convergent sequence $x := (\xi_k)$ with $\varsigma := \lim_k \xi_k$, such that $v_k/\lambda_k = \xi_k - \varsigma$. Thus, we have proved that, for every sequence $(v_k) \in m$ there exists a sequence $(\xi_k) \in m^\lambda$ such that $v_k = \lambda_k(\xi_k - \varsigma)$. Hence, $A_\lambda \in (m, c)$. This implies (by Exercise 1.8) that condition (8.3) is satisfied,

$$\lim_n \sum_k \frac{|a_{nk} - \delta_k|}{\lambda_k} = 0 \tag{8.6}$$

and

$$\phi := \lim_n A_n x = \sum_k \frac{\delta_k}{\lambda_k} v_k + \varsigma \lim_n \mathfrak{A}_n.$$

If $\mu_n = O(1)$, then it is necessary to replace $O(1)$ by $o(1)$ in (8.4); that is equivalent to (8.6).

If $\mu_n \neq O(1)$, then writing

$$\mu_n(A_n x - \phi) = \mu_n \sum_k \frac{a_{nk} - \delta_k}{\lambda_k} v_k + \varsigma \mu_n(\mathfrak{A}_n - \lim_n \mathfrak{A}_n), \tag{8.7}$$

we conclude, by (8.2), that the method $A_{\lambda,\mu} \in (m, m)$, where

$$A_{\lambda,\mu} := \left(\mu_n \frac{a_{nk} - \delta_k}{\lambda_k} \right).$$

Hence, using Exercise 1.6, we conclude that condition (8.4) is satisfied.

If $\lambda_n = O(1)$, then the proof is similar to the case $\lambda_n \neq O(1)$, but now $v_k = o(1)$, and, instead of Exercise 1.8, it is necessary to use Exercise 1.3.

Sufficiency. Let conditions (1.2) and (8.2)–(8.4) be fulfilled. Then, relation (8.5) also holds for every $x \in m^\lambda$ and $(\mathfrak{A}_n) \in m^\mu$ by (8.2). If $\lambda_n \neq O(1)$ and $\mu_n = O(1)$, then $A_\lambda \in (m, c)$ by (1.2), (8.3), and (8.6) (remembering that, in this case, instead of (8.4) we have (8.6)); that is, $A \in (m^\lambda, c)$.

If $\lambda_n \neq O(1)$ and $\mu_n \neq O(1)$, then the validity of (8.6) follows from the validity of (8.4). In that case also $A_\lambda \in (m, c)$ by (1.2), (8.3), and (8.6). Moreover, using Exercise 1.8, we obtain that the limit ϕ exists and is finite, and therefore relation (8.7) holds for every $x \in m^\lambda$. Hence, $A_{\lambda,\mu} \in (m, m)$ by (8.4) (see Exercise 1.6), and, consequently, $A \in (m^\lambda, m^\mu)$ by (8.2). The proof for the case $\lambda_n = O(1)$ is analogous. $\square$

Definition 8.4 We say that a method A preserves the λ-boundedness if $A \in (m^\lambda, m^\lambda)$.

As $\delta_k \equiv 0$ for an Sq–Sq regular method A by Theorem 1.1, from Theorem 8.1 we obtain, for $\mu = \lambda$, the following result.

Corollary 8.1 Let $A = (a_{nk})$ be an Sq–Sq regular method with $\mathfrak{A}_n = 1$. Then, A preserves the λ-boundedness if and only if

$$\lambda_n \sum_k \frac{|a_{nk}|}{\lambda_k} = O(1). \tag{8.8}$$

Example 8.1 Let $A = (a_{nk})$ be the Zweier method; that is, $A = Z_{1/2}$. It is easy to see that $Z_{1/2}$ is Sq–Sq regular with $\mathfrak{A}_n = 1$ for $n \geq 1$. In addition,

$$\lambda_n \sum_k \frac{|a_{nk}|}{\lambda_k} = \frac{1}{2}\left(\frac{\lambda_n}{\lambda_{n-1}} + 1\right).$$

Therefore, $Z_{1/2}$ preserves the λ-boundedness if and only if $\lambda_n/\lambda_{n-1} = O(1)$.

Theorem 8.2 A method $A = (a_{nk}) \in (c^\lambda, c^\mu)$ if and only if conditions (8.3) and (8.4) are fulfilled and

$$Ae^k \in c^\mu, \tag{8.9}$$

$$Ae \in c^\mu, \tag{8.10}$$

$$A\lambda^{-1} \in c^\mu; \; \lambda^{-1} := \left(\frac{1}{\lambda_k}\right). \tag{8.11}$$

If $A \in (c^\lambda, c^\mu)$, then

$$\lim_n \mu_n(A_n x - \phi) = \sum_k a_k^{\lambda,\mu}(v_k - v) + \lim_n \mu_n(\mathfrak{A}_n - \delta)\varsigma$$

$$+ \lim_n \mu_n\left(\sum_k \frac{u_{nk}}{\lambda_k} - a^\lambda\right)v, \tag{8.12}$$

where $\phi := \lim_n A_n x$, $v := \lim_k v_k$ and

$$\delta := \lim_n \mathfrak{A}_n, \delta_k := \lim_n a_{nk}, a^\lambda := \lim_n \sum_k \frac{a_{nk}}{\lambda_k}; \; a_k^{\lambda,\mu} := \lim_n \mu_n \frac{a_{nk} - \delta_k}{\lambda_k}.$$

Proof: **Necessity.** Assume that $A \in (c^\lambda, c^\mu)$ and prove that conditions (8.3), (8.4), and (8.9)–(8.11) are fulfilled. It is easy to see that $e^k \in c^\lambda$, $e \in c^\lambda$, and $\lambda^{-1} \in c^\lambda$. Hence, conditions (8.9)–(8.11) hold. As equality (8.5) holds for every $x := (\xi_k) \in c^\lambda$, and the finite limit δ exists by (8.10), then the method A_λ transforms this convergent sequence (v_k) into c. Similar to the proof of the necessity of Theorem 8.1, it is possible to show that, for every sequence $(v_k) \in c$, there exists a sequence $(\xi_k) \in c^\lambda$ such that $v_k = \lambda_k(\xi_k - \varsigma)$. Hence, $A_\lambda \in (c, c)$. This implies (by Exercise 1.1) that the finite limits δ_k and a^λ

exist, and that condition (8.3) is satisfied. Using (8.5), for every $x \in c^\lambda$, we can write

$$\phi := \lim_n A_n x = a^\lambda v + \sum_k \frac{\delta_k}{\lambda_k}(v_k - v) + \delta \varsigma, \tag{8.13}$$

where $\varsigma := \lim_k \xi_k$ and $v := \lim_k v_k$. Now, using (8.5) and (8.13), we obtain

$$\mu_n(A_n x - \phi) = \mu_n \sum_k \frac{a_{nk} - \delta_k}{\lambda_k}(v_k - v) + \mu_n(\mathfrak{A}_n - \delta)\varsigma$$

$$+ \mu_n \left(\sum_k \frac{a_{nk}}{\lambda_k} - a^\lambda \right) v. \tag{8.14}$$

The finite limits for the two last summands in the right side of (8.14) in the process $n \to \infty$ exist by (8.10) and (8.11). This implies that the method $A_{\lambda,\mu} \in (c_0, c)$. Therefore, using Exercise 1.3, we can conclude that condition (8.4) holds. Finally, relation (8.12) follows from (8.14).

Sufficiency. Suppose that conditions (8.3), (8.4), and (8.9)–(8.11) are fulfilled. First we note that relation (8.14) holds for every $x \in c^\lambda$ and the finite limits δ_k, δ and a^λ exist correspondingly by (8.9), (8.10), and (8.11). As (8.3) also holds, then $A_\lambda \in (c, c)$ by virtue of Exercise 1.1, and therefore relations (8.13) and (8.14) hold for every $x \in c^\lambda$. Now the finite limits for the two last summands in the right side of (8.14) in the process $n \to \infty$ exist by (8.9) and (8.11). In addition, using conditions (8.4) and (8.9), we can assert, using Exercise 1.3, that $A_{\lambda,\mu} \in (c_0, c)$. Thus, $A \in (c^\lambda, c^\mu)$. □

It is easy to see that conditions (1.2) and (8.3) imply condition

$$\sum_k \frac{|\delta_k|}{\lambda_k} < \infty. \tag{8.15}$$

Conditions (8.4) and (8.15) imply condition (8.3). Therefore, from Theorems 8.1 and 8.2, we get the following corollary.

Corollary 8.2 Condition (8.3) in Theorems 8.1 and 8.2 can be replaced by condition (8.15).

Remark 8.1 For $\mu_n = \lambda_n \equiv 1$, from Theorem 8.1 or 8.2, we get the Kojima–Schur theorem (or the result of Exercise 1.1).

Definition 8.5 A method A is said to be λ-conservative if $A \in (c^\lambda, c^\lambda)$.

If $\lambda_k = O(1)$, then the notion of λ-conservativity coincides with the notion of conservativity. Hence, we assume that $\lambda_k \neq O(1)$. It appears that the

λ-conservative method A can be characterized with the help of the method $A_{\lambda,\mu}$ with $\mu = \lambda$; that is, by the method $A_{\lambda,\lambda}$.

Theorem 8.3 A method $A = (a_{nk})$ is λ-conservative if and only if condition (8.3) is fulfilled, $Ae \in c^\lambda$ and the method $A_{\lambda,\lambda}$ is conservative.

Proof: From the proof of Theorem 8.2 we can conclude, using Exercise 1.1, that, if conditions (8.3), (8.4), (8.9), and (8.10) hold for $\mu = \lambda$, then it is sufficient to prove that condition (8.11) is equivalent to the existence of the finite limit

$$\lim_n \mu_n \sum_k \frac{a_{nk} - \delta_k}{\lambda_k}. \tag{8.16}$$

First, we see that condition (8.15) is valid by (8.3) and (8.9). This implies that $(\delta_k/\lambda_k) \in cs$, and due to (8.4) and the assumption $\lambda_k \neq O(1)$, we obtain that

$$\left| \sum_k \frac{a_{nk}}{\lambda_k} - \sum_k \frac{\delta_k}{\lambda_k} \right| \leq \sum_k \frac{|a_{nk} - \delta_k|}{\lambda_k},$$

and relation (8.6) holds. Consequently,

$$\lim_n \sum_k \frac{a_{nk}}{\lambda_k} = \sum_k \frac{\delta_k}{\lambda_k},$$

from which it follows that condition (8.11) is equivalent to the existence of the finite limit (8.16). ☐

Example 8.2 Let $A = (a_{nk})$ be the Zweier method; that is, $A = Z_{1/2}$. As $\delta_k \equiv 0$, then $A_{\lambda,\lambda} = (a_{nk}^\lambda)$ is a lower triangular matrix with

$$a_{n,n-1}^\lambda = \frac{1}{2} \frac{\lambda_n}{\lambda_{n-1}}, a_{nn}^\lambda = \frac{1}{2} \text{ and } a_{nk}^\lambda = 0 \text{ for } k < n - 1.$$

Hence,

$$\lim_n a_{nk}^\lambda = 0 \text{ for every } k$$

and

$$\sum_{k=0}^{n} a_{nk}^\lambda = \frac{1}{2} \left(\frac{\lambda_n}{\lambda_{n-1}} + 1 \right).$$

Therefore, using Exercise 1.1, we conclude that $Z_{1/2}$ is λ-conservative if and only if the limit $\lim_n (\lambda_n/\lambda_{n-1})$ exists.

Definition 8.6 We say that a method A improves the λ-convergence, if $A \in (c^\lambda, c^\mu)$ with $\lim_n \mu_n/\lambda_n = \infty$.

Lemma 8.1 If a method A improves the λ-convergence, then

$$\lim_n \sum_{k=0}^{n} |a_{nk} - \delta_k| = 0. \tag{8.17}$$

Proof: Suppose, by contradiction, that condition (8.17) does not hold. Then, there exists a number $\epsilon > 0$ and indexes i_n, such that

$$\sum_{k=0}^{i_n} |a_{i_n k} - \delta_k| \geq \epsilon; \ n = 0, 1, \dots .$$

Hence,

$$\sum_{k=0}^{i_n} \frac{|a_{i_n k} - \delta_k|}{\lambda_k} \geq \frac{\epsilon}{\lambda_{i_n}}.$$

In addition, $A \in (c^\lambda, c^\mu)$, since A improves the λ-convergence. Therefore, using (8.4), we obtain

$$\frac{\epsilon}{\lambda_{i_n}} = O\left(\frac{1}{\mu_{i_n}}\right).$$

However, this relation contradicts the condition $\lim_n \mu_n/\lambda_n = \infty$. Thus, equality (8.17) holds. □

Using Lemma 8.1 and Theorem 1.1, we immediately get the following corollary.

Corollary 8.3 An Sq–Sq regular method cannot improve the λ-convergence.

Finally, we note that Theorems 8.1 and 8.2 and Corollary 8.3 were first proved by Kangro [17] and [19]. Theorem 8.3 is proved in [24]. From [17, 19], and [24], the reader can find more results on matrix transforms of c^λ and m^λ. The improvement of the λ-convergence has been studied in [4].

8.3 Matrix Transforms from m_A^λ into m_B^μ

In this section, we describe necessary and sufficient conditions for $M \in (m_A^\lambda, m_B^\mu)$, if $A = (a_{nk})$ is a normal method, $B = (b_{nk})$ is a triangular method, $M = (m_{nk})$ is an arbitrary matrix, and $\lambda := (\lambda_k), \mu := (\mu_k)$, monotonically increasing sequences with $\lambda_k > 0, \mu_k > 0$.

Let $A^{-1} := (\eta_{nk})$ be the inverse matrix of a normal method A. Then,

$$\sum_{k=0}^{j} m_{nk}\xi_k = \sum_{k=0}^{j} m_{nk} \sum_{l=0}^{k} \eta_{kl}y_l = \sum_{l=0}^{j} h_{jl}^n$$

for each $x := (\xi_k) \in m_A^\lambda$, where $y_l := A_l x$ and $H^n := (h_{jl}^n)$ is the lower triangular matrix for every fixed n, with

$$h_{jl}^n := \sum_{k=l}^{j} m_{nk}\eta_{kl}, \, l \leq j.$$

This implies that the transformation $y = Mx$ exists for every $x \in m_A^\lambda$ if and only if the matrix $H^n := (h_{jl}^n) \in (m^\lambda, c)$ for every fixed n. Consequently, using Exercise 8.4, we immediately obtain the following result.

Proposition 8.1 Let $A = (a_{nk})$ be a normal method and $M = (m_{nk})$ an arbitrary matrix. Then, the transformation $y = Mx$ exists for every $x \in m_A^\lambda$ if and only if condition (5.14) is satisfied and

$$\lim_j \sum_{l=0}^{j} h_{jl}^n \text{ exists and is finite,} \tag{8.18}$$

$$\sum_l \frac{|h_{jl}^n|}{\lambda_l} = O_n(1) \tag{8.19}$$

and

$$\lim_j \sum_{l=0}^{j} \frac{|h_{jl}^n - h_{nl}|}{\lambda_l} = 0. \tag{8.20}$$

Also, condition (8.19) can be replaced by the condition

$$\sum_l \frac{|h_{nl}|}{\lambda_l} = O_n(1). \tag{8.21}$$

Now we are able to prove the main result of this section

Theorem 8.4 Let $A = (a_{nk})$ be a normal method, $B = (b_{nk})$ a triangular method, and $M = (m_{nk})$ an arbitrary matrix. Then, $M \in (m_A^\lambda, m_B^\mu)$ if and only if conditions (5.14), (5.25), (8.18)–(8.20) are satisfied and

$$\sum_l \frac{|\gamma_{nl}|}{\lambda_l} = O(1), \tag{8.22}$$

$$\mu_n \sum_l \frac{|\gamma_{nl} - \gamma_l|}{\lambda_l} = O(1), \tag{8.23}$$

$$(\rho_n) \in m^\mu, \rho_n := \lim_j \sum_{l=0}^j \gamma_{nl}^j, \tag{8.24}$$

where $\Gamma^n := (\gamma_{nl}^j)$ is the lower triangular matrix for every fixed n with

$$\gamma_{nl}^j := \sum_{k=l}^j g_{nk}\eta_{kl}, l \le j.$$

Also, condition (8.22) can be replaced by the condition

$$\sum_l \frac{|\gamma_l|}{\lambda_l} < \infty, \tag{8.25}$$

and, if $\mu_n = O(1)$ and $\lambda_n \ne O(1)$, then it is necessary to replace the $O(1)$ in (8.23) by $o(1)$.

Proof: **Necessity.** Assume that $M \in (m_A^\lambda, m_B^\mu)$. Then, the transformation $y = Mx$ exists for every $x \in m_A^\lambda$. Hence, conditions (5.14) and (8.18)–(8.20) hold by Proposition 8.1, and equation (5.27) holds for every $x \in m_A^\lambda$ because the change of the order of summation is allowed by the triangularity of B. From (5.27), we can conclude that $G := BM \in (m_A^\lambda, m^\mu)$. In addition,

$$\sum_{k=0}^j g_{nk}\xi_k = \sum_{l=0}^j \gamma_{nl}^j A_l x \tag{8.26}$$

for every $x \in m_A^\lambda$. By the normality of A, there exists an $x \in m_A^\lambda$, such that $(A_l x) = e$. Consequently, condition (8.24) is satisfied by (8.26).

Assume now that $\lambda_n \ne O(1)$. Then, by the normality of A, for each bounded sequence (β_n) there exists an $x \in m_A^\lambda$, such that

$$\lim_n A_n x := \phi \text{ and } \beta_n = \lambda_n(A_n x - \phi) \tag{8.27}$$

(see also the proof of the necessity of Theorem 8.1). Moreover, using (8.26) and (8.27), we obtain

$$\sum_{k=0}^j g_{nk}\xi_k = \phi \sum_{l=0}^j \gamma_{nl}^j + \sum_{l=0}^j \frac{\gamma_{nl}^j}{\lambda_l} \beta_l \tag{8.28}$$

for every $x \in m_A^\lambda$. As the series $G_n x$ are convergent for every $x \in m_A^\lambda$, and the finite limits ρ_n exist by (8.24), then the matrix $\Gamma_\lambda^n := (\gamma_{nl}^j/\lambda_l) \in (m, c)$ for every n. Therefore, from (8.24), we obtain, using Exercise 1.8, that

$$G_n x = \phi \rho_n + \sum_l \frac{\gamma_{nl}}{\lambda_l} \beta_l \tag{8.29}$$

for every $x \in m_A^\lambda$. In addition, the finite limit $\lim_n \rho_n := \rho$ exists by (8.24). Therefore, from (8.29), we can conclude that the matrix $\Gamma_\lambda := (\gamma_{nl}/\lambda_l) \in (m, c)$. Consequently, conditions (5.25) and (8.22) hold,

$$\lim_n \sum_l \frac{|\gamma_{nl} - \gamma_l|}{\lambda_l} = 0, \tag{8.30}$$

and

$$\lim_n G_n x = \phi \rho_n + \sum_l \frac{\gamma_l}{\lambda_l} \beta_l \tag{8.31}$$

for every $x \in m_A^\lambda$ by Exercise 1.8. Now it is clear that, for $\mu_n = O(1)$, it is necessary to replace $O(1)$ in (8.23) by $o(1)$; that is, condition (8.23) is equivalent to (8.30).

We continue with the case $\mu_n \neq O(1)$, writing

$$\mu_n(G_n x - \lim_n G_n x) = \phi \mu_n(\rho_n - \rho) + \mu_n \sum_l \frac{\gamma_{nl} - \gamma_l}{\lambda_l} \beta_l \tag{8.32}$$

for every $x \in m_A^\lambda$. This implies that the matrix $\Gamma_{\lambda,\mu} := (\mu_n(\gamma_{nl} - \gamma_l)/\lambda_l) \in (m, m)$. Hence, using Exercise 1.6, we conclude that condition (8.23) holds.

If $\lambda_n = O(1)$, then the proof is similar to the case $\lambda_n \neq O(1)$, but now $\beta_l = o(1)$, and, instead of Exercise 1.8, it is necessary to use Exercise 1.3.

Finally, we note that the necessity of condition (8.25) follows from the validity of conditions (8.22) and (8.23).

Sufficiency. Let all of the conditions of the present theorem be fulfilled. Then, the transformation $y = Mx$ exists for every $x \in m_A^\lambda$ by Proposition 8.1, and equations (5.27) and (8.26)–(8.28) hold for every $x \in m_A^\lambda$. As in the proof of the necessity of the present theorem, we get, using (8.24) and Exercise 1.8, that, from (8.28), follows the validity of (8.29) for every $x \in m_A^\lambda$. If $\lambda_n \neq O(1)$ and $\mu_n = O(1)$, then $\Gamma_\lambda^n \in (m, c)$ for every n by (5.25), (8.22), and (8.30) (in this case, instead of (8.23), we have (8.30)); that is, $M \in (m_A^\lambda, c_B)$.

If $\lambda_n \neq O(1)$ and $\mu_n \neq O(1)$, then the validity of (8.30) follows from the validity of (8.23). Thus, in that case, again $\Gamma_\lambda \in (m, c)$ by (5.25), (8.22), and (8.30). Moreover, relation (8.31) holds for every $x \in m_A^\lambda$ by virtue of Exercise 1.8, and therefore relation (8.32) holds for every $x \in m^\lambda$. Hence, $\Gamma_{\lambda,\mu} \in (m, m)$ by (8.23) (see Exercise 1.6) and, consequently, $M \in (m_A^\lambda, m_B^\mu)$ by (8.24).

The proof for the case $\lambda_n = O(1)$ is analogous.

Condition (8.22) can be replaced by (8.25) because the validity of (8.22) follows from the validity of (8.23) and (8.25). □

Remark 8.2 If M is a matrix with finite rows, then conditions (5.14) and (8.18)–(8.20) are redundant in Theorem 8.4.

We note that for $\lambda_n = O(1)$ and $\mu_n = O(1)$, Theorem 8.4 gives necessary and sufficient conditions for $M \in (c_A, c_B)$ for a normal method A and a triangular Method B. For a matrix, given in the form (5.1), from Theorem 8.4 we obtain Theorem 20.2 of [24] concerning summability factors from m_A^λ into m_B^λ. If $A = B = E$ (E the identity method), then Theorem 8.4 reduces to Theorem 8.1.

Now we consider necessary and sufficient conditions for A and B to be M^{seq}-consistent on m_A^λ.

Theorem 8.5 Let $A = (a_{nk})$ be a normal method, $B = (b_{nk})$ a triangular method and $M = (m_{nk})$ an arbitrary matrix. Then, A and B are M^{seq}-consistent on m_A^λ if and only if conditions (5.14), (8.18)–(8.20) are satisfied and

$$\lim_n \rho_n = 1, \tag{8.33}$$

$$\lim_n \sum_l \frac{|\gamma_{nl}|}{\lambda_l} = 0. \tag{8.34}$$

Proof: **Necessity.** Assume that A and B are M^{seq}-consistent on m_A^λ. Then, $M \in (m_A^\lambda, c_B)$ and therefore conditions (5.14), (8.18)–(8.20) are satisfied, relation (8.29) holds for every $x \in m_A^\lambda$ and the finite limit ρ_n exist (for each n), and the finite limit $\lim_n \rho_n = \rho$ (see the proof of the necessity of Theorem 8.4). We show now that $\rho = 1$. Indeed, there exists an $x \in m_A^\lambda$, such that $(A_l x) = e$, since A is normal. For this x, we have

$$\lim_n G_n x = \lim_n A_n x = 1,$$

and $\rho = 1$ by (8.26); that is, condition (8.33) holds. From (8.29), we can conclude that $\Gamma_\lambda = (\gamma_{nl}/\lambda_l) \in (m, c_0)$. Hence, using Exercise 1.9, we obtain that condition (8.34) holds.

Sufficiency. Let all of the conditions of the present theorem be satisfied. Then, the finite limits ρ_n exist and relation (8.29) holds for every $x \in m_A^\lambda$ (see the proof of Theorem 8.4). Hence, A and B are M^{seq}-consistent on m_A^λ because $\Gamma_\lambda \in (m, c_0)$. □

Remark 8.3 We note that, for the method (R, p_n), the preserving of the λ-boundedness means that $(R, p_n) \in (ms^\lambda, m^\lambda)$, where

$$ms^\lambda := \{x = (\xi_k) \mid (X_n) \in m^\lambda\}; \quad X_n := \sum_{k=0}^n \xi_k.$$

Using Theorem 8.4, we prove the following result.

Theorem 8.6   Let $B = (b_{nk})$ be a triangular method, $M = (m_{nk})$ an arbitrary matrix, and (R, p_n) a method preserving the λ-boundedness and satisfying the properties

$$P_n = 0(P_{n-1}), \tag{8.35}$$

$$\frac{P_n}{p_n} = 0\left(\frac{P_{n+1}}{p_{n+1}}\right). \tag{8.36}$$

Then, $M \in (m_{(R,p_n)}^\lambda, m_B^\mu)$ if and only if condition (6.34) holds and

$$e^0 \in m_G^\mu, \tag{8.37}$$

$$\sum_l \frac{1}{\lambda_l}\left|P_l\Delta_l\frac{\Delta_l m_{nl}}{p_l}\right| = O_n(1), \tag{8.38}$$

$$\lim_j \frac{P_j m_{nj}}{p_j \lambda_j} = 0, \tag{8.39}$$

$$\sum_l \frac{1}{\lambda_l}\left|P_l\Delta_l\frac{\Delta_l g_{nl}}{p_l}\right| = O(1), \tag{8.40}$$

$$\mu_n \sum_l \frac{1}{\lambda_l}\left|P_l\Delta_l\frac{\Delta_l(g_{nl} - g_l)}{p_l}\right| = O(1). \tag{8.41}$$

In addition, condition (8.40) can be replaced by the condition

$$\sum_l \frac{1}{\lambda_l}\left|P_l\Delta_l\frac{\Delta_l g_l}{p_l}\right| < \infty. \tag{8.42}$$

Proof: **Necessity.** Assume that $M \in (m_{(R,p_n)}^\lambda, m_B^\mu)$. As in Proposition 7.1, we see that equalities (7.13) and (7.14) hold. This implies that condition (8.38) holds, by Theorem 8.4. Now it is not difficult to see that

$$\sum_l \frac{|h_{jl}^n - h_{nl}|}{\lambda_l} = \left|\frac{P_{j-1}m_{n,j+1}}{p_j\lambda_{j-1}}\right| + \left|\frac{P_j m_{nj}}{p_j\lambda_j} - \frac{P_j}{\lambda_j}\Delta_j\frac{\Delta_j m_{nj}}{p_j}\right|. \tag{8.43}$$

In addition, condition (8.38) implies that

$$\lim_j \frac{P_j}{\lambda_j}\Delta_j\frac{\Delta_j m_{nj}}{p_j} = 0. \tag{8.44}$$

Hence, condition (8.39) holds by Theorem 8.4.

As $e^k \in ms^\lambda$ and $ms^\lambda \subset m^\lambda_{(R,p_n)}$ by assumption (see Remark 8.3), then condition (6.34) holds. Since $a_{n0} \equiv 1$ for the method (R,p_n), condition (8.37) is satisfied due to Exercise 8.6. As relation (7.84) holds, then conditions (8.40)–(8.42) hold by Theorem 8.4.

Sufficiency. Suppose that all of the conditions of the present theorem are satisfied. First, conditions (5.14) and (8.19) hold by (7.13), (7.14), and (8.38). Now relation (8.44) holds by (8.38). We show that condition (8.20) also holds. Indeed, using Exercise 8.7, we can conclude that

$$\frac{P_{j-1}}{\lambda_{j-1}} = O(1)\frac{P_{j+1}}{\lambda_{j+1}}.$$

Consequently, using (8.35), (8.36), and (8.39), we can write

$$\frac{P_{j-1}m_{n,j+1}}{p_j\lambda_{j-1}} = O(1)\frac{P_{j+1}m_{n,j+1}}{p_j\lambda_{j+1}} = O(1)\frac{P_j m_{n,j+1}}{p_j\lambda_{j+1}} = O(1)\frac{P_{j+1}m_{n,j+1}}{p_{j+1}\lambda_{j+1}} = o_n(1).$$

Therefore, condition (8.43) implies (8.20) by (8.39) and (8.44). Finally, from conditions (6.34), (8.40), and (8.41) follow the validity of (5.25), (8.22), and (8.23). Thus, $M \in (m^\lambda_{(R,p_n)}, m^\mu_B)$, due to Theorem 8.4 and Exercise 8.6.

Also, it is easy to see that conditions (8.41) and (8.42) imply (8.40). □

Using Theorems 8.5 and 8.6 and Exercise 8.6, we immediately get the following corollary.

Corollary 8.4   Let $B = (b_{nk})$ be a triangular method, $M = (m_{nk})$ an arbitrary matrix, and (R,p_n) a method preserving the λ-boundedness, satisfying the properties (8.35) and (8.36). Then, (R,p_n) and B are M^{seq}-consistent on $m^\lambda_{(R,p_n)}$ if and only if conditions (8.38) and (8.39) are satisfied, and

$$\lim_n g_{n0} = 1, \tag{8.45}$$

$$\lim_n \sum_l \frac{1}{\lambda_l}\left|P_l\Delta_l\frac{\Delta_l g_{nl}}{p_l}\right| = 0. \tag{8.46}$$

Remark 8.4   If M is a matrix with finite rows, then conditions (8.38) and (8.39) are redundant in Theorem 8.6 and in Corollary 8.4.

We note that Theorems 8.4–8.6 were first proved in [2]. For the summability factors from $m^\lambda_{(R,p_n)}$ into m^μ_B, necessary and sufficient conditions were found in [19].

8.4 On Orders of Approximation of Fourier Expansions

As an application of the results on the convergence and summability with speed, in this section, the comparison of the orders of approximation of Fourier expansions in Banach spaces by different matrix methods has been considered. For this purpose, we use a lower triangular matrix $M^f = (m_{nk})$, generated by a continuous function f on $[0, \infty[$, where $f(0) = 1$ and $f(t) = 0$, if $t \geq 1$, that is,

$$m_{nk} = f(k \kappa_n), k \leq n,$$

where (κ_n) is a sequence satisfying

$$\kappa_n > 0, \kappa_n = O\left(\frac{1}{n+1}\right). \tag{8.47}$$

Also, we use the series-to-sequence Zygmund method Z^r of order r $(r > 0)$, defined by the lower triangular matrix $A = (a_{nk})$, where

$$a_{nk} := 1 - \left(\frac{k}{n+1}\right)^r, k \leq n. \tag{8.48}$$

It is easy to see that the Riesz method $(R, p_n) = Z^r$ if $p_n = (n+1)^r - n^r$.

Let X be a Banach space with norm $\| \cdot \|$, and $c(X)$, $cs(X)$, and $c_A(X)$ be, respectively, the spaces of convergent sequences, convergent series, and A-summable sequences in X. Also we define the following sequence spaces:

$$m^\lambda(X) := \{x = (\xi_k) \mid \xi_k \in X, \exists \lim \xi_k = \zeta, \lambda_k \parallel \xi_k - \zeta \parallel = O(1)\},$$

$$ms^\lambda(X) := \left\{ x = (\xi_k) \mid \xi_k \in X, (X_n) \in m^\lambda(X), X_n = \sum_{k=0}^n \xi_k \right\},$$

$$m_A^\lambda(X) := \{x = (\xi_k) \mid \xi_k \in X, \exists \lim_n A_n x = v, \lambda_n \parallel A_n x - v \parallel = O(1)\}.$$

Remark 8.5 All of the results of Chapters 5–7 hold if scalar-valued sequences, or sequence sets, are replaced by the corresponding X-valued sequences or sequence spaces ([19], p. 139).

Using Remark 8.5, we can use the results from Section 8.3 to obtain the comparison of the orders of approximation of Fourier expansions in Banach spaces by different matrix methods. We now describe the notion of a Fourier expansion. Suppose that a total sequence of mutually orthogonal continuous projections (T_k) $(k = 0, 1, \ldots)$ on X exists; that is, T_k is a bounded linear operator of X into itself, $T_k x = 0$ for all k implies $x = 0$, and $T_j T_k = \delta_{jk} T_k$. Then, we may associate to each x from X, a formal Fourier expansion

$$x \sim \sum_k T_k x.$$

We know (see [39], pp. 74–75, 85–86) that the sequence of projections (T_k) exists if, for example, $X = C_{2\pi}$ is the set of all 2π-periodic functions, which are uniformly continuous and bounded on $\mathbf{R}$, and $X = L^p_{2\pi}$ $(1 \leq p < \infty)$ is the set of all 2π-periodic functions, Lebesgue integrable to the p-th power over $(-\pi, \pi)$ or $X = L^p(\mathbf{R})$ $(1 \leq p < \infty)$ is the set of all functions, Lebesgue integrable to the p-th power over $\mathbf{R}$. A good treatment of Fourier expansions can be found, for example, in [12] or [39]. If $X = C_{2\pi}$ or $X = L^p_{2\pi}$ $(1 \leq p < \infty)$, then it is known (see [12], p. 106) that, for the classical trigonometric system (T_k) and $0 < \alpha < 1$, the relation

$$(n+1)^\alpha \parallel Z^1_n x - x \parallel = O_x(1)$$

holds if and only if

$$x \in Lip\ \alpha := \{x \in X|\ \parallel x(t+h) - x(t) \parallel = O_x(h^\alpha)\}.$$

Several results of this type are known, where the order of approximation can be characterized via the Lipschitz conditions (see [12], pp. 67–88, 106–107). In this section, we compare the relation between the orders of approximation of Fourier expansions by Z^r and M^f.

Theorem 8.7 Let X be a Banach space, $x \in X, \alpha > 0, \beta > 0$, and $0 < \gamma \leq 1$, where conditions

$$\beta - \gamma + 1 \leq \alpha < r, \tag{8.49}$$

$$u(t) := t^{1-r} f'(t) \in Lip\ \gamma \tag{8.50}$$

are satisfied. Then, the relation

$$(n+1)^\alpha \parallel Z^r_n x - x \parallel = O_x(1) \tag{8.51}$$

implies that

$$(n+1)^\beta \parallel M^f_n x - x \parallel = O_x(1). \tag{8.52}$$

Proof: Let $\lambda_n := (n+1)^\alpha, \mu_n := (n+1)^\beta$ and $\lambda := (\lambda_n), \mu := (\mu_n)$. Then, $\lambda_n \neq O(1)$ and $\mu_n \neq O(1)$. Hence, it is sufficient to show that the conditions of Theorem 8.6 and Corollary 8.4 hold for $(R, p_n) = Z^r, M = M^f$ and $B = (\delta_{nk})$. As $(R, p_n) = Z^r$, if $p_n = (n+1)^r - n^r$, then it is easy to see that conditions (8.35) and (8.36) are satisfied. For the series-to-sequence Zigmund method defined by (8.48), we obtain, by (5.9), that the corresponding sequence-to-sequence Zygmund method is presented by $\tilde{A} = (\tilde{a}_{nk})$, where

$$\tilde{a}_{nk} = \Delta_k a_{nk} = \frac{(k+1)^r - k^r}{(n+1)^r}.$$

Hence,

$$\mathfrak{A}_n = \sum_{k=0}^{n} \tilde{a}_{nk} = \sum_{k=0}^{n} \frac{(k+1)^r - k^r}{(n+1)^r} = 1.$$

Moreover,

$$(n+1)^\alpha \sum_{k=0}^{n} \frac{|\tilde{a}_{nk}|}{(k+1)^\alpha} = (n+1)^{\alpha-r} \sum_{k=0}^{n} \frac{(k+1)^r - k^r}{(k+1)^\alpha}$$

$$= O(1)(n+1)^{\alpha-r} \sum_{k=0}^{n} (k+1)^{r-\alpha-1} = O(1),$$

by the mean-value theorem of Lagrange. Hence, the method Z^r preserves the λ-boundedness by Corollary 8.1.

As $g_{nk} = m_{nk} = f(k\kappa_n)$, then $g_{n0} \equiv 1$ and $\lim_n g_{nk} = f(0) = 1$. Thus, conditions (6.34), (8.37), (8.40), and (8.45) are satisfied. We now show that conditions (8.41) and (8.46) also hold. Indeed,

$$L := \mu_n \sum_{l=0}^{n} \frac{1}{\lambda_l} \left| P_l \Delta_l \frac{\Delta_l(g_{nl} - g_l)}{p_l} \right| = \mu_n \sum_{l=0}^{n} \frac{1}{\lambda_l} \left| P_l \Delta_l \frac{\Delta_l g_{nl}}{p_l} \right| = L_1 + L_2 + L_3,$$

where

$$L_1 := \mu_n \sum_{l=0}^{n-2} \frac{1}{\lambda_l} \left| P_l \Delta_l \frac{\Delta_l g_{nl}}{p_l} \right|$$

$$= (n+1)^\beta \kappa_n^{\ r} \sum_{l=0}^{n-2} (l+1)^{r-\alpha} \left| \Delta_l \frac{f((l+1)\kappa_n) - f(l\kappa_n)}{(\kappa_n(l+1))^r - (\kappa_n l)^r} \right|,$$

$$L_2 := \frac{\mu_n}{\lambda_l} \left| P_l \Delta_l \frac{\Delta_l g_{nl}}{p_l} \right| \bigg|_{l=n-1}$$

$$= (n+1)^\beta (\kappa_n)^r n^{r-\alpha} \left| \Delta_l \frac{f(n\kappa_n) - f((n-1)\kappa_n)}{(n\kappa_n)^r - ((n-1)\kappa_n)^r} \right|,$$

$$L_3 := \frac{\mu_n}{\lambda_n} \left| \frac{P_n g_{nn}}{p_n} \right| = (n+1)^{\beta-\alpha+r} (\kappa_n)^r \left| \frac{f((n+1)\kappa_n) - f(n\kappa_n)}{((n+1)\kappa_n)^r - (n\kappa_n)^r} \right|.$$

Consequently, we have

$$L_1 = \frac{1}{r}(n+1)^\beta O((n+1)^{-r}) \sum_{l=0}^{n-2} (l+1)^{r-\alpha} |\Delta_l[(\kappa_n(l+\theta_l))^{1-r} f'(\kappa_n(l+\theta_l))]|$$

$$= O(1)(n+1)^{\beta-r-\gamma} \sum_{l=0}^{n-2} (l+1)^{r-\alpha} = O(1)(n+1)^{\beta-\alpha-\gamma+1} = O(1),$$

$$L_2 = \frac{1}{r}(n+1)^\beta O((n+1)^{-r}) n^{r-\alpha} |\Delta_l[(\kappa_n(n-1+\theta_1))^{1-r} f'(\kappa_n(n-1+\theta_1))]|$$

$$= O(1)(n+1)^{\beta-\alpha-\gamma} = O(1),$$

$$L_3 = \frac{1}{r}(n+1)^{\beta-\alpha+r} O((n+1)^{-r})(\kappa_n(n+\theta_2))^{1-r} |f'(\kappa_n(n+\theta_2))|$$

$$= O(1)(n+1)^{\beta-\alpha-\gamma} = O(1)$$

$(0 < \theta_l, \theta_1, \theta_2 < 1)$, by the mean-value theorem of Cauchy, and conditions (8.47), (8.49), and (8.50), since, from (8.49), we have $\beta - r - \gamma \leq 0$. Therefore, $L = O(1)$; that is, condition (8.41) is satisfied. As the sequence μ is unbounded, then condition (8.46) also holds. As conditions (8.38) and (8.39) are redundant in Theorem 8.6 and Corollary 8.4 by Remark 8.4, then all of the conditions of Theorem 8.6 and Corollary 8.4 hold. Thus, relation (8.52) holds by Theorem 8.6 and Corollary 8.4. □

It is easy to see that Theorem 8.7 can be reformulated as follows.

Theorem 8.8   Let X be a Banach space. If the function u, defined by $u(t) :=$ $t^{1-r} f'(t)$ on $]0, 1[$, belongs to *Lip* γ, where $\gamma \in]0, 1]$, and the estimation (8.51) holds for some $x \in X$ and, for $\alpha \in]1 - \gamma, r[$, then

$$(n + 1)^{\alpha + \gamma - 1} \parallel M_n^f x - x \parallel = O_x(1). \tag{8.53}$$

We note that for $\kappa_n = 1/(n + 1)$, Theorem 8.7 was proved in [1] and for $f(t) = 1 - t^s$ with $s > 0$ (then M^f is the Zygmund method Z^s) in [2].

Now define functions f_i $(i = 1, \dots, 6)$ on $[0, 1]$ as follows:

$$f_1(t) = \cos(\pi t/2);$$

$$f_2(t) = (1 - t^\omega)^\sigma \quad (\omega, \sigma > 0);$$

$$f_3(t) = \begin{cases} 1 - 6t^2 + 6t^3 & (t \in [0, \tfrac{1}{2}]), \\ 2(1 - t)^3 & (t \in [\tfrac{1}{2}, 1]); \end{cases}$$

$$f_4(t) = (1 - t)\cos(\pi t) + \frac{1}{\pi}\sin(\pi t);$$

$$f_5(t) = 1 - \tan^2(\frac{\pi t}{4});$$

$$f_6(t) = \begin{cases} 1 & (t = 0), \\ \frac{\pi t}{2}\cot(\frac{\pi t}{2}) & (t \in]0, 1]). \end{cases}$$

In the following examples, we apply Theorems 8.7 and 8.8 in the cases when $M^\varphi = M^{\varphi_i}$ $(i = 1, \dots, 6)$. The method M^{φ_1} is called the method of Rogozinski ([40], p. 284), M^{φ_2} the method of Riesz ([12], p. 265, 475), M^{φ_3} the method of Jackson–de La Vallée Poussin ([12], p. 205), M^{φ_4} the method of Bohman–Korovkin ([14], p. 305), M^{φ_5} the method of Zhuk ([40], p. 319), and M^{φ_6} the method of Favard ([23], p. 161).

We shall use the notation

$$u_i(t) := t^{1-r} f_i'(t); \quad t \in]0, 1[, i = 1, \dots, 6.$$

Example 8.3   Let X be a Banach space, $x \in X, \alpha > 0, \beta > 0$, and at least one of the conditions

$$\beta \leq \alpha < r \leq 1, \tag{8.54}$$

$$1 < r < 2 \text{ and } \beta + r - 1 \leq \alpha < r, \tag{8.55}$$

$$\beta \leq \alpha < r = 2 \tag{8.56}$$

be satisfied. We shall show that relation (8.51) implies (8.52) for $f = f_1$. To accomplish this, we shall prove that all of the conditions of Theorem 8.7 are satisfied for $f = f_1$, if at least one of the conditions (8.54)–(8.56) is fulfilled. As

$$f_1'(t) = -\frac{\pi}{2} \sin\left(\frac{\pi t}{2}\right) \ (t \in [0, 1]),$$

then

$$u_1(t) = -\left(\frac{\pi}{2}\right)^r y^{1-r} \sin y, \text{ where } y := \frac{\pi t}{2} \ (t \in [0, 1]).$$

Hence, $u_1 \in Lip\ 1$ on $]0, 1[$, if u_1' is bounded on $[0, 1]$, or h', where $h(y) = y^{1-r} \sin y$, is bounded on $[0, \pi/2]$. As

$$h'(y) = (1 - r)y^{-r} \sin y + y^{1-r} \cos y = y^{1-r}\left((1 - r)\frac{\sin y}{y} + \cos y\right),$$

then h' is bounded on $[0, \pi/2]$, if $r \leq 1$. Also h' is bounded on $[0, \pi/2]$, if $r = 2$, since, in this case,

$$h(y) = \frac{\sin y}{y}; \ \lim_{y \to 0+} h'(y) = \lim_{y \to 0+} \frac{y \cos y - \sin y}{y^2} = 0.$$

Consequently, if condition (8.54) or (8.56) is fulfilled, we can take $\gamma = 1$ in conditions (8.49) and (8.50).

Let now $1 < r < 2$. Then, $h \in Lip\ 1$ on every segment $[a, \pi/2]$, where $1 < a < \pi/2$. Since

$$h(y) = y^{2-r}\frac{\sin y}{y},$$

then $h(y)$ is equivalent to y^{2-r} in the limit process $y \to 0+$. Consequently, $h \in Lip\ (2 - r)$ on $[0, \pi/2]$. This implies that $u_1 \in Lip\ (2 - r)$ on $[0, 1]$, so we can take $\gamma = 2 - r$ in conditions (8.49) and (8.50). Thus, relation (8.51) implies (8.52), if at least one of the conditions (8.54)–(8.56) is satisfied by Theorem 8.7.

Example 8.4 Let X be a Banach space. Assume that, for some $x \in X$, and for $\alpha \in]0, r[$, estimate (8.51) holds. We prove that, then estimate (8.53) holds for $f = f_2$, if at least one of the following conditions is fulfilled:

$$\gamma = 1, \sigma \geq 2 \text{ and } \omega \geq r + 1 \text{ or } \omega = r \geq 1, \tag{8.57}$$

$$\max\{0, 1 - \alpha\} < \gamma = \omega - r < 1 \text{ and } \sigma \geq 2, \tag{8.58}$$

$$\max\{0, 1 - \alpha\} < \gamma = \sigma - 1 < 1 \text{ and } \omega > r + 1 \text{ or } \omega = r \geq 1, \tag{8.59}$$

$$\max\{0, 1 - \alpha\} < \gamma = \min\{\omega - r, \sigma - 1\} \text{ and } \max\{\omega - r, \sigma - 1\} < 1, \tag{8.60}$$

$$\max\{0, 1 - \alpha\} < \gamma = \omega = r < 1 \text{ and } \sigma \geq 2, \tag{8.61}$$

$$\max\{0, 1 - \alpha\} < \gamma = \min\{\omega, \sigma - 1\}, \max\{\omega, \sigma - 1\} < 1 \text{ and } \omega = r. \tag{8.62}$$

Assume that condition (8.51) holds. It is sufficient to show that, at least one condition, from conditions (8.57)–(8.62), implies the validity of the conditions of Theorem 8.8 for $f = f_2$. As the method of the proof for all cases (8.57)–(8.62) is quite similar, we give a proof of this theorem only for conditions (8.57), (8.59), and (8.62).

Let condition (8.57) be fulfilled. As in this case

$$u_2(t) = -\sigma\omega(1 - t^\omega)^{\sigma-1}t^{\omega-r} \ (t \in]0, 1[)$$

and

$$u_2'(t) = -\sigma\omega t^{\omega-r-1}(1 - t^\omega)^{\sigma-2}[(\omega - r)(1 - t\omega) - (\sigma - 1)\omega t^\omega] \ (t \in]0, 1[),$$

then u_2' is bounded on $]0, 1[$ for $\omega \geq r + 1$. Also, u_2' is bounded on $]0, 1[$ for $\omega = r$ since, in that case,

$$u_2'(t) = \sigma(\sigma - 1)\omega^2 t^{\omega-1}(1 - t^\omega)^{\sigma-2} \ (t \in]0, 1[).$$

Hence, $u_2 \in Lip\ 1$ on $[0, 1]$. Thus, all of the conditions of Theorem 8.8 hold.

Let condition (8.59) be fulfilled. Then, due to $\omega \geq 1$, the derivative u_2' is bounded on $]0, a]$ for each $a \in]0, 1[$. Hence, $u_2 \in Lip\ 1$ on $]0, a]$. In addition, $u_2 \in Lip\ (\sigma - 1)$ on $]a, 1[$ since u_2 is equivalent to $-\sigma\omega(1 - t^\omega)^{\sigma-1}$ in the limit process $t \to 1-$ for $\omega > r + 1$, and

$$u_2(t) = -\sigma\omega(1 - t^\omega)^{\sigma-1} \ (t \in]0, 1[) \tag{8.63}$$

for $\omega = r$. This implies that $u_2 \in Lip\ (\sigma - 1)$ on $]0, 1[$. Thus, the conditions of Theorem 8.8 hold.

Let condition (8.62) be satisfied. Then, using (8.63), we obtain $u_2 \in Lip\ (\sigma - 1)$ on $]a, 1[$ for each $a \in]0, 1[$. Moreover, $u_2 \in Lip\ \omega$ on $]0, a]$ because $t^\omega \in Lip\ \omega$ on $]0, a]$ if $\omega \in]0, 1[$. Consequently, $u_2 \in Lip\ (\min\{\omega, \sigma - 1\})$ on $]0, 1[$. Thus, the conditions of Theorem 8.8 hold.

Example 8.5 Let X be a Banach space. Assume that, for some $x \in X$ and, for $\alpha \in]0, r[$, estimate (8.51) holds. We prove that, in this case, estimate (8.53) holds for $f = f_i$ with $i = 3, 4, 5$, if at least one of the following conditions is satisfied:

$$\gamma = 1 \text{ and } r \leq 1, \tag{8.64}$$

$$\max\{0, 1 - \alpha\} < \gamma = 2 - r < 1, \tag{8.65}$$

$$\gamma = 1 \text{ and } r = 2. \tag{8.66}$$

We shall show that estimate (8.53) holds for $f = f_6$, if condition (8.64) or (8.65) is satisfied.

First, we assume that estimate (8.51) is fulfilled. It is sufficient to show, that at least one condition from (8.64) to (8.66) implies the conditions of Theorem 8.8 for suitable $f = f_i$. We give a proof only for condition (8.65) if $i = 3, 4$, since the other cases are similar. Suppose that condition (8.65) holds for $i = 3$ and $i = 4$. Then,

$$u_3(t) = \begin{cases} -6t^{2-r}(2 - 3t) & (t \in]0, \frac{1}{2}]), \\ -6t^{1-r}(1 - t)^2 & (t \in]\frac{1}{2}, 1[) \end{cases}$$

and

$$u_4(t) = \pi(t - 1)t^{1-r} \sin(\pi t)(t \in]0, 1[).$$

As

$$u_3'(t) = \begin{cases} -6t^{1-r}[(2 - r)(2 - 3t) - 3t] & (t \in]0, \frac{1}{2}]), \\ -6t^{-r}[(1 - r)(1 - t)^2 - 2(1 - t)t] & (t \in]\frac{1}{2}, 1[) \end{cases}$$

and

$$u_4'(t) = \pi t^{1-r} \left[\pi(t + (1 - r)(t - 1))\frac{\sin(\pi t)}{\pi t} + \pi(t - 1)\cos(\pi t) \right] (t \in]0, 1[),$$

the derivatives u_3' and u_4' are bounded on $[a, 1[$ for every $a \in]0, 1[$. Consequently, $u_3, u_4 \in Lip\ 1$ on $[a, 1[$ for every $a \in]0, 1[$. Moreover, in the limit process $t \to 0+$, the function u_3 is equivalent to $-12t^{2-r}$ and u_4 to $-\pi^2 t^{2-r}$, since u_4 can be presented in the form

$$u_4 = \pi^2(t - 1)t^{2-r}\frac{\sin(\pi t)}{\pi t} \quad (t \in]0, 1[),$$

Therefore, $u_3, u_4 \in Lip\ (2 - r)$ on $]0, 1[$, and so the conditions of Theorem 8.8 hold.

Theorem 8.7 and Example 8.3 first was proved in [1], and Examples 8.4 and 8.5 in [3]. Also, convergence and summability with speed have been used for estimating the order of approximation of Fourier and orthogonal series by several authors (see, e.g., [5–9, 20–22, 29–31]).

8.5 Exercise

Exercise 8.1 Prove that the Riesz method $(\tilde{R}, p_n)$ preserves the λ-boundedness if and only if

$$\frac{\lambda_n}{|P_n|} \sum_{k=0}^{n} \frac{|p_k|}{\lambda_k} = O(1).$$

Hint. Use Corollary 8.1.

Exercise 8.2   Prove that, if $p_k > 0$ for every k, $\lim_k P_k = \infty$, and $\lambda_k \neq O(1)$, then $(\tilde{R}, p_n)$ is λ-conservative if and only if there exists the finite limit

$$\lim_n \frac{\lambda_n}{P_n} \sum_{k=0}^{n} \frac{p_k}{\lambda_k}.$$

Hint. Show that, for $A = (\tilde{R}, p_n)$ with $p_k > 0$ and $\lim_k P_k = \infty$, the method $(\tilde{R}, p_n)$ is λ-conservative if and only if the matrix

$$(\tilde{R}, p_n)_{\lambda,\lambda} := \left(\frac{\lambda_n p_k}{P_n \lambda_k} \right)$$

is conservative.

Exercise 8.3   Prove that $A = (a_{nk}) \in (c^\lambda, m^\mu)$ if and only if $A \in (m^\lambda, m^\mu)$.

Hint. Find necessary and sufficient conditions for $A_{\lambda,\mu} \in (c_0, m)$; further, use Exercise 1.6.

Exercise 8.4   Prove that a matrix $A = (a_{nk}) \in (m^\lambda, c)$ if and only if conditions (1.2), (8.3), and (8.6) are satisfied, and the finite limit $\lim_n \mathfrak{A}_n := \delta$ exists. Show that condition (8.3) can be replaced by condition (8.15).

Hint. Use Theorem 8.1 and Corollary 8.2.

Exercise 8.5   Prove that, if a method B is normal, then condition (8.19) is redundant in Theorem 8.4.

Hint. Show that condition (8.19) follows from (8.22).

Exercise 8.6   Prove that if, a method $A = (a_{nk})$ has the property that $a_{n0} \equiv 1$, then, in Theorem 8.4, condition (8.18) is redundant, and condition (8.24) can be replaced by condition (8.37).

Hint. Show that

$$\sum_{l=0}^{j} h_{jl}^n = m_{n0} \text{ and } \sum_{l=0}^{j} \gamma_{nl}^j = g_{n0}.$$

Then, use Theorem 8.4, Remark 5.1, and Lemma 7.3.

Exercise 8.7   Prove that if the Riesz method preserves λ-boundedness, then

$$\frac{\lambda_n P_l}{\lambda_l P_n} = O(1) \ (l \leq n).$$

Hint. Use Exercise 8.1.

Exercise 8.8   Prove that (8.51) implies (8.53) for $f = f_2$ in Example 8.4, if at least one of the conditions (8.58), (8.60), or (8.61) is fulfilled.

Hint. Show that all of the conditions of Theorem 8.8 hold.

Exercise 8.9   Prove that (8.51) implies (8.53) for $f = f_5$ in Example 8.5, if condition (8.65) is fulfilled.

Exercise 8.10   Prove that (8.51) implies (8.53) for $f = f_i$ with $i = 3, 4$ in Example 8.5, if condition (8.64) or (8.66) is fulfilled.

Exercise 8.11   Prove that (8.51) implies (8.53) for $f = f_6$ in Example 8.5, if condition (8.64) or (8.65) is fulfilled.

References

1 Aasma, A.: Comparison of orders of approximation of Fourier expansions by different matrix methods. Facta Univ. Ser. Math. Inform. **12**, 233–238 (1997).

2 Aasma, A.: Matrix transformations of λ-boundedness fields of normal matrix methods. Stud. Sci. Math. Hung. **35**(1-2), 53–64 (1999).

3 Aasma, A.: On the summability of Fourier expansions in Banach spaces. Proc. Estonian Acad. Sci. Phys. Math. **51**(3), 131–136 (2002).

4 Аасма, А.: Convergence acceleration and improvement by regular matrices. In: Dutta, H. and Rhoades, B.E. (eds.) Current Topics in Summability Theory and Applications, pp. 141–180. Springer, Singapore (2016).

5 Aljanǎić, S.: The modulus of continuity of Fourier series transformed by convex multipliers (Serbo-Croatian). Acad. Serbe Sci. Arts Glas. **254**, 35–53 (1963).

6 Aljanǎić, S.: Über konvexe Multiplikatoren bei Fourier-Reihen (German). Math. Z. **81**, 215–222 (1963).

7 Aljanǎić, S.: The modulus of continuity of Fourier series transformed by convex multipliers. II. (Serbian). Acad. Serbe Sci. Arts Glas. **260**, 99–105 (1965).

8 Aljanǎić, S. and Tomić, M.: Über den Stetigkeitsmodul von Fourier-Reihen mit monotonen Koeffizienten. (German). Math. Z. **88**, 274–284 (1965).

9 Aljanǎić, S. and Tomić, M.: Transformationen von Fourier-Reihen durch monoton abnehmende Multiplikatoren. Acta Math. Acad. Sci. Hung. **17**, 23–30 (1966).

10 Brezinski, C.: Convergence acceleration during the 20th century. J. Comput. Appl. Math. **122**(1-2), 1–21 (2000).

11 Brezinski, C. and Redivo-Zaglia, M.: Extrapolation Methods. Theory and Practice. North-Holland, Amsterdam (1991).

12 Butzer, P.L. and Nessel, R.I.: Fourier Analysis and Approximation: One-Dimensional Theory. Birkhäuser Verlag, Basel and Stuttgart (1971).

13 Caliceti, E.; Meyer-Hermann, M.; Ribeca, P.; Surzhykov, A. and Jentschura, U.D.: From useful algorithms for slowly convergent series to physical predictions based on divergent perturbative expansions. Phys. Rep. **446**(1-3), 1–96 (2007).

14 Higgins, J.R: Sampling Theory in Fourier and Signal Analysis: Foundations. Clarendon Press, Oxford (1996).

15 Jürimäe, E.: Properties of matrix mappings on rate-spaces and spaces with speed. Tartu Ül. Toimetised **970**, 53–64 (1994).

16 Jürimäe, E.: Matrix mappings between rate-spaces and spaces with speed. Tartu Ül. Toimetised **970**, 29–52 (1994).

17 Kangro, G.: O množitelyah summirujemosti tipa Bora-Hardy dlya zadannoi ckorosti I (On the summability factors of the Bohr-Hardy type for a given speed I). Eesti NSV Tead. Akad. Toimetised Füüs.-Mat. **18**(2), 137–146 (1969).

18 Kangro, G.: O množitelyah summirujemosti tipa Bora-Hardy dlya zadannoi ckorosti II (On the summability factors of the Bohr-Hardy type for a given speed II). Eesti NSV Tead. Akad. Toimetised Füüs.-Mat. **18**(4), 387–395 (1969).

19 Kangro, G.: Množiteli summirujemosti dlya ryadov, λ-ogranitšennõh metodami Rica i Cezaro (Summability factors for the series λ-bounded by the methods of Riesz and Cesàro). Tartu Riikl. Ül. Toimetised **277**, 136–154 (1971).

20 Kangro, G.: Poryadok summirovanya ortogonalnyh ryadov treugolnymi regulyarnymi metodami II. (The rate of summability of orthogonal series by triangular regular methods. II.). Eesti NSV Tead. Akad. Toimetised Füüs.-Mat. **23**, 107–112 (1974).

21 Kangro, G.: Poryadok summirovanya ortogonalnyh ryadov treugolnymi regulyarnymi metodami I. (The rate of summability of orthogonal series by triangular regular methods. I.). Eesti NSV Tead. Akad. Toimetised Füüs.-Mat. **23**, 3–11 (1974).

22 Kangro, G.: Silnaja summirujemost ortogonalnyh ryadov so skorostju. (The strong summability of orthogonal series with speed). Eesti NSV Tead. Akad. Toimetised Füüs.-Mat. **28**, 1–8 (1979).

23 Korneichuk, N.P.: Exact Constants in Approximation Theory. Cambridge University Press, Cambridge (1991).

24 Leiger, T.: Funktsionaalanalüüsi meetodid summeeruvusteoorias (Methods of functional analysis in summability theory). Tartu Ülikool, Tartu (1992).

25 Leiger, T. and Maasik, M.: O λ-vklyutchenyy matriz summirovaniya (The λ-inclusion of summation matrices). Tartu Riikl. Ül. Toimetised **770**, 61–68 (1987).

26 Šeletski, A. and Tali, A.: Comparison of speeds of convergence in Riesz-Type families of summability methods. Proc. Est. Acad. Sci. **57**(1), 70–80 (2008).

27 Šeletski, A. and Tali, A.: Comparison of speeds of convergence in Riesz-Type families of summability methods II. Math. Model. Anal. **15**(1), 103–112 (2010).

28 Sidi, A.: Practical Extrapolation Methods, Cambridge Monographs on Applied and Computational Mathematics, Vol. 10. Cambridge University Press, Cambridge (2003).

29 Sikk, J.: Multiplikatorŏ, T^λ-dopolnitelnŏe prostranstva i koeffitsientŏ Fourier dlya nekotorŏh klassov funktsii (Multipliers, T^λ-complementary spaces, and the Fourier coefficients of certain classes of functions). Tartu Riikl. Ül. Toimetised **374**, 180–185 (1975).

30 Sikk, J.: Nekatorŏe, T^λ-konstruktivnŏe prostranstva i multiplikatorŏ dliya klassov X_{T^λ}, X_{U^μ} (Certain T^λ-constructive spaces and multipliers of classes X_{T^λ}, X_{U^μ}). Tartu Riikl. Ül. Toimetised **374**, 163–179 (1975).

31 Sikk, J.: Dopolnitelnŏe prostranstva koeffitsientov Fourier so skorostju (Complementary spaces of Fourier coefficients with a rate). Tartu Riikl. Ül. Toimetised **355**, 222–235 (1975).

32 Sikk, J.: Matrix mappings for rate-space and $\mathcal{K}$-multipliers in the theory of summability. Tartu Riikl. Ül. Toimetised **846**, 118–129 (1989).

33 Sikk, J.: The rate-spaces $m(\lambda)$, $c(\lambda)$, $c_0(\lambda)$ and $l^p(\lambda)$ of sequences. Tartu Ül. Toimetised **970**, 87–96 (1994).

34 Stadtmüller, U. and Tali, A.: Comparison of certain summability methods by speeds of convergence. Anal. Math. **29**(3), 227–242 (2003).

35 Tammeraid, I.: Generalized linear methods and convergence acceleration. Math. Model. Anal. **8**(1), 87–92 (2003).

36 Tammeraid, I.: Convergence acceleration and linear methods. Math. Model. Anal. **8**(4), 329–335 (2003).

37 Tammeraid, I.: Several remarks on acceleration of convergence using generalized linear methods of summability. J. Comput. Appl. Math. **159**(2), 365–373 (2003).

38 Tammeraid, I.: Generalized Riesz method and convergence acceleration. Math. Model. Anal. **9**(4), 341–348 (2004).

39 Trebels, W.: Multipliers for (C, α)-bounded Fourier Expansions in Banach Spaces and Approximation Theory, Lecture Notes in Mathematics, Vol. 329. Springer-Verlag, Berlin-Heidelberg, New York (1973).

40 Zhuk, V.V.: Approksimatsiya periodicheskih funktsii (Approximation of periodic functions). Leningrad State University, Leningrad (1982).

9

On Convergence and Summability with Speed II

9.1 Introduction

All notions and notations not defined in this chapter can be found in Chapters 5–8. In this chapter, we continue the study of convergence, boundedness, and summability with speed, begun in Chapter 8. Throughout the present chapter, we suppose that $\lambda = (\lambda_k)$ (called the speed of convergence or summability) is a sequence with $0 < \lambda_k \nearrow \infty$ if not specified otherwise. The notions of λ-reversible and λ-perfect matrix methods are introduced.

Let A be a matrix with real or complex entries. In Section 9.2, we present some topological properties of the spaces m^λ, c^λ, c_A^λ, and m_A^λ, which have been introduced, for example, in [2, 3, 5–6, 8–9], and [11].

Let M be a matrix with real or complex entries. In Section 9.3, necessary and sufficient conditions for $M \in (c_A^\lambda, m_B^\mu)$, $M \in (c_A^\lambda, c_B^\mu)$, and for the M^{seq}-consistency of A and B on c_A^λ (μ is another speed and B is another matrix method) are described. As an application of the main results, the matrix transforms for the cases of Riesz and Cesàro methods are investigated.

Necessary and sufficient conditions for $M \in (c_A^\lambda, c_B^\mu)$ and for $M \in (c_A^\lambda, m_B^\mu)$ were first proved in [1] for a λ-reversible (this notion is defined in Section 9.2) or a λ perfect method A (this notion is defined in Section 9.3) and a triangular method B. If a matrix M is of the form (5.1), where (ε_k) is a sequence of numbers, then the above-mentioned problem reduces to the problem of finding necessary and sufficient conditions for numbers ε_k to be the B^μ-summability or B^μ-boundedness factors for c_A^λ. We say that numbers ε_k are the B^μ-summability factors for c_A^λ if $(\varepsilon_k x_k) \in c_B^\mu$ for every $x := (x_k) \in c_A^\lambda$. Similarly, B^μ-boundedness factors can be defined for c_A^λ. If B is the identity method, that is, $B = E$, and

An Introductory Course in Summability Theory, First Edition. Ants Aasma, Hemen Dutta, and P.N. Natarajan.
© 2017 John Wiley & Sons, Inc. Published 2017 by John Wiley & Sons, Inc.

$\mu_k = O(1)$, then we obtain necessary and sufficient conditions for the numbers ε_k to be convergence factors for c_A^λ. Necessary and sufficient conditions for numbers ε_k to be the B^μ-boundedness or B^μ-summability factors for c_A^λ were found by Kangro [7, 10]. If, in addition, $\varepsilon_k \equiv 1$, then $m_{nk} = \delta_{nk}$, that is, $Mx = x$ for each $x \in c_A^\lambda$. So, we get the inclusion problems $c_A^\lambda \subset c_B^\mu$ and $c_A^\lambda \subset m_B^\mu$, which were first studied by Leiger and Maasik (see [12]).

9.2 Some Topological Properties of m^λ, c^λ, c_A^λ and m_A^λ

In this section, we introduce some topological properties of the spaces m^λ, c^λ, c_A^λ, and m_A^λ.

Proposition 9.1 m^λ and c^λ are BK-spaces with respect to the norm

$$\|x\|_\lambda := \sup_n \{|v_k|, |\varsigma|\}; \ x := (\xi_k),$$

where

$$\varsigma := \lim_k \xi_k, v_k = \lambda_k(\xi_k - \varsigma).$$

Proof: It is easy to check the validity of the norm's axioms for $\|x\|_\lambda$ on m^λ and we leave it to the reader. Let us show that m^λ is complete with respect to this norm. Let (x^n) be a fundamental sequence in m^λ, where $x^n := (\xi_k^n)$, that is,

$$\lim_{n,j} \|x^n - x^j\|_\lambda = 0. \tag{9.1}$$

Denoting

$$\xi^n := \lim_k \xi_k^n, v_k^n := \lambda_k(\xi_k^n - \xi^n), \tag{9.2}$$

we obtain, by (9.1), that

$$\lim_{n,j} \|v^n - v^j\|_m = 0; \ \lim_{n,j} |\xi^n - \xi^j| = 0$$

($\|\cdot\|_m$ denotes the norm in m) uniformly with respect to k, where $v^n := (v_k^n) \in m$. Thus, v^n and (ξ^n) are fundamental sequences, respectively, in the Banach spaces m and $\mathbf{K}$, where $\mathbf{K} = \mathbf{C}$ or $\mathbf{K} = \mathbf{R}$. Hence, the limits

$$\xi := \lim_n \xi^n \in \mathbf{K}, v = (v_k) := \lim_n v^n \in m \tag{9.3}$$

exist and are finite. This implies $\lim_n v_k^n = v_k$, since m is a BK-space. Therefore, from the second relation of (9.2), we get

$$\lim_n \xi_k^n = \lim_n \frac{1}{\lambda_k}(v_k^n + \xi^n) = \frac{1}{\lambda_k}(v_k + \xi) := x_k.$$

Let $x := (x_k)$. As $v \in m$ and $1/\lambda_k \to 0$ if $k \to \infty$, then $x \in c$. In addition,

$$\xi = \lim_k x_k; \ v_k = \lambda_k(x_k - \xi).$$

Consequently, $x \in m^\lambda$, and $\lim_n \|x^n - x\|_\lambda = 0$ by (9.2), that is, $x^n \to x$ if $n \to \infty$ in m^λ. Thus, m^λ is a Banach space. Moreover, as

$$|\lim x| \le \|x\|_\lambda \text{ and } |v_k| \le \|x\|_\lambda$$

for $x \in c^\lambda$, then the linear functionals $\lim$ and $x \to v_k$ are continuous in m^λ. Therefore, using the equality

$$x_k = \frac{1}{\lambda_k}(v_k + \lim x),$$

we obtain that the coordinate functionals are continuous for every k. Thus, m^λ is a BK-space.

Now we prove that c^λ is a closed subset of m^λ. Let (x^n) be a sequence of elements of c^λ converging to x in the BK-space m^λ. Thus, in this case

$$v^n := (v^n_k) \in c; \ v = (v_k) \in m$$

and

$$\lim_n \|v^n - v\|_m = 0.$$

As c is closed in m, then $v = (v_k) \in c$, and hence, $x \in c^\lambda$. It means that c^λ is a closed subset of m^λ, and therefore c^λ is also a BK space. $\qquad\square$

Proposition 9.2 Every element $x := (\xi_k) \in c^\lambda$ can be represented in the form

$$x - \varsigma e \mid v\lambda^{-1} + \sum_k \frac{v_k - v}{\lambda_k} e^k, \tag{9.4}$$

where

$$\varsigma := \lim_k \xi_k, \ \lambda^{-1} := (1/\lambda_k),$$

and

$$v_k = \lambda_k(\xi_k - \varsigma), v := \lim_k v_k.$$

Proof: For each $x := (\xi_k) \in c^\lambda$ we denote

$$x^n := \varsigma e + v\lambda^{-1} + \sum_{k=0}^n \frac{v_k - v}{\lambda_k} e^k.$$

Then, we obtain

$$\|x - x^n\|_\lambda = \sup_{k>n} |v_k - v| = 0$$

by the definition of $\|\cdot\|_\lambda$. Hence, relation (9.4) holds. $\qquad\square$

Proposition 9.3 The general form of a continuous linear functional $f \in (c^\lambda)'$ can be presented by the formula

$$f(x) = s\varsigma + tv + \sum_k \tau_k v_k, (\tau_k) \in l, s, t \in \mathbf{C}. \tag{9.5}$$

Proof: Applying the functional $f \in (c^\lambda)'$ to both sides of (9.4), and denoting

$$s := f(e), u := f(\lambda^{-1}) \text{ and } \tau_k := \frac{f(e^k)}{\lambda_k},$$

we can write

$$f(x) = s\varsigma + uv + \sum_k \tau_k(v_k - v), x \in c^\lambda. \tag{9.6}$$

We shall show that $(\tau_k) \in l$. From (9.6), we get relation (9.5) with

$$t := u - \sum_k \tau_k.$$

We determine the elements ξ_k^n of x^n by the equalities

$$\xi_k^n := \begin{cases} sgn \ \tau_k / \lambda_k & (k \le n), \\ 0 & (k > n). \end{cases}$$

Then, $\lim_k \xi_k^n = 0$,

$$v_k^n := \begin{cases} sgn \ \tau_k & (k \le n), \\ 0 & (k > n), \end{cases}$$

$v(x^n) = 0$ and $\|x^n\|_\lambda \le 1$. Therefore,

$$\|f\| \ge \sup_n |f(x^n)| = \sup_n \sum_{k=0}^n |\tau_k| = \sum_k |\tau_k|,$$

which implies that $(\tau_k) \in l$. So we proved that every $f \in (c^\lambda)'$ can be represented in form (9.5). It is easy to verify that every functional, represented in form (9.5), is continuous and linear on c^λ. $\qquad\square$

Definition 9.1 A method $A = (a_{nk})$ is called λ-reversible, if the infinite system of equations $z_n = A_n x$ has a unique solution, for each sequence $(z_n) \in c^\lambda$.

For a λ-reversible matrix A, the following result holds.

Proposition 9.4 c_A^λ for a λ-reversible method A is a BK-space with respect to the norm

$$\|x\|_{A,\lambda} := \|Ax\|_\lambda = \sup_n \{|\beta_n|, |\phi|\}; \ x := (\xi_k),$$

where

$$\phi := \lim_n A_n x, \beta_n = \lambda_n(A_n x - \phi).$$

The general form of a continuous linear functional $f \in (c_A^\lambda)'$ can be represented by the formula

$$f(x) = s\phi + t\beta + \sum_n \tau_n \beta_n, (\tau_n) \in l, s, t \in \mathbf{C}, \tag{9.7}$$

where

$$\beta := \lim_n \beta_n.$$

Proof: Using Propositions 9.1–9.3, the present proposition has a proof similar to the proof of Lemma 5.3. □

Propositions 9.1–9.4 were first proved in [9] (see also [11]). References [2, 3, 5, 8, 9, 11] contain other results on the structure and topological properties of m^λ, c^λ, c_A^λ, and m_A^λ.

Using Proposition 9.4, we can prove the following result.

Corollary 9.1 Let $A = (a_{nk})$ be a λ-reversible method. Then, every coordinate ξ_k of a sequence $x := (\xi_k) \in c_A^\lambda$ can be represented in the form

$$\xi_k = \phi \eta_k + \beta \varphi_k + \sum_n \frac{\eta_{kn}}{\lambda_n}(\beta_n - \beta), \tag{9.8}$$

where $\eta := (\eta_k)$, $\varphi := (\varphi_k)$ and $\eta := (\eta_{kj})$, for each fixed j are the solutions of the system $y = Ax$ corresponding to $y = (\delta_{nn})$, $y = (\delta_{nn}/\lambda_n)$ and $y = (y_n) = (\delta_{nj})$, and $(\eta_{kn}/\lambda_n) \in l$, for every fixed k.

Proof: As c_A^λ is a BK-space by Proposition 9.4, every coordinate ξ_k of a sequence $x := (\xi_k) \in c_A^\lambda$ is a continuous linear functional on c_A^λ. Hence, by Proposition 9.4, there exist numbers η_k, t_k, and a sequence $(\tau_{kn}) \in l$ for every fixed k, such that

$$\xi_k = \phi \eta_k + \beta t_k + \sum_n \tau_{kn} \beta_n. \tag{9.9}$$

If now $A_n x = \delta_{nj}$ for every fixed j, then, from (9.9), we obtain $\xi_k = \tau_{kj}\lambda_j$ for every fixed j since

$$\lim_n \delta_{nj} = 0, \beta_n = \lambda_n \delta_{nj}$$

and

$$\lim_n \lambda_n \delta_{nj} = 0.$$

Thus, the sequence $\eta := (\eta_{kj})$ for every fixed j, with $\eta_{kj} = \tau_{kj}\lambda_j$, is a solution of the system of equations $y = Ax$ for $y = (y_n) = (\delta_{nj})$. For $A_n x = \delta_{nn}$, from (9.9), we get $\xi_k = \eta_k$. In this case

$$\lim_n \delta_{nn} = 1, \beta_n = 0 \text{ and } \lim_n \beta_n = 0.$$

Hence, $\eta := (\eta_k)$ is the solution of the system $y = Ax$ for $y = (\delta_{nn})$. If $y = (\delta_{nn}/\lambda_n)$, then, from (9.9), we obtain

$$\xi_k = t_k + \sum_n \tau_{kn} = t_k + \sum_n \frac{\eta_{kn}}{\lambda_n},$$

since

$$\lim_n \frac{\delta_{nn}}{\lambda_n} = 0, \beta_n = \delta_{nn} = 1 \text{ and } \lim_n \beta_n = 1.$$

Hence, $\varphi := (\varphi_k)$ with

$$\varphi_k = t_k + \sum_n \frac{\eta_{kn}}{\lambda_n},$$

is the solution of the system $y = Ax$ for $y = (\delta_{nn}/\lambda_n)$. Moreover, $(\tau_{kn}) = (\eta_{kn}/\lambda_n) \in l$ for every fixed k. □

Remark 9.1 It is easy to see that every reversible (and also every normal) method A is λ-reversible since $c^\lambda \subset c$. Therefore, for a reversible method A, relation (9.8) reduces to relation (5.5) (with $x_k \equiv \xi_k$) and, for a normal method A, relation (9.8) reduces to the relation

$$\xi_k = \sum_{l=0}^k \eta_{kl} A_l x$$

for every $x \in c_A^\lambda$, due to (5.7).

Example 9.1 If A is the sequence-to-sequence Riesz method, that is, $A = (\tilde{R}, p_n)$, then relation (9.8) takes the same form, due to Remark 9.1, as in Example 5.1.

9.3 Matrix Transforms from c_A^λ into c_B^μ or m_B^μ

In this section, we describe necessary and sufficient conditions for $M \in (c_A^\lambda, c_B^\mu)$ and for $M \in (c_A^\lambda, m_B^\mu)$, if $A = (a_{nk})$ is a λ-reversible method, $B = (b_{nk})$ a triangular matrix, $M = (m_{nk})$ and $\lambda := (\lambda_k), \mu := (\mu_k)$ are monotonically increasing sequences with $\lambda_k > 0$, $\mu_k > 0$. First we introduce (and also remember) some essential notations, which we shall use throughout in this section. We use $\eta := (\eta_k), \varphi := (\varphi_k)$ and (η_{kl}) (for every fixed l) to denote the solutions of the system

$y = Ax$, respectively, for $y = (\delta_{nn})$, $y = (\delta_{nn}/\lambda_n)$ and $y = (y_n) = (\delta_{nl})$. Then, we set

$$\gamma := \lim_n G_n \eta; \quad \psi := \lim_n G_n \varphi.$$

For every $x \in c_A^\lambda$, we define

$$\phi := \lim_n A_n x; \quad \beta_n = \lambda_n(A_n x - \phi); \quad \beta := \lim_n \beta_n.$$

We shall now find necessary and sufficient conditions for the transformation $y = Mx$ to exist for every $x \in c_A^\lambda$.

Proposition 9.5 Let $A = (a_{nk})$ be a λ-reversible method and $M = (m_{nk})$ an arbitrary matrix. Then, the transformation $y = Mx$ exists for every $x \in c_A^\lambda$ if and only if conditions (5.14), (5.15), and (8.19) are satisfied and

$$(m_{nk}\varphi_k) \in cs \text{ for every fixed } n. \tag{9.10}$$

Proof: **Necessity.** Assume that the transformation $y = Mx$ exists for every $x \in c_A^\lambda$. By Corollary 9.1, every coordinate ξ_k of a sequence $x := (\xi_k) \in c_A^\lambda$ can be represented in the form (9.8), where $(\eta_{kn}/\lambda_n) \in l$ for every fixed k. Hence, we can write

$$\sum_{k=0}^{j} m_{nk}\xi_k = \phi \sum_{k=0}^{j} m_{nk}\eta_k + \beta \sum_{k=0}^{j} m_{nk}\varphi_k + \sum_l \frac{h_{jl}^n}{\lambda_l}(\beta_l - \beta) \tag{9.11}$$

for every sequence $x := (\xi_k) \in c_A^\lambda$. It is easy to see that $\eta \in c_A^\lambda$ and $\varphi \in c_A^\lambda$ since $e \in c_A^\lambda$, $\lambda^{-1} \in c_A^\lambda$, and A is λ-reversible. Consequently, conditions (5.15) and (9.10) are satisfied. Using (9.11), we obtain that the matrix $H_\lambda^n := (h_{jl}^n/\lambda_l)$ for every n transforms the sequence $(\beta_l - \beta) \in c_0$ into c. Using the λ-reversibility of A, it is possible to show (see the proof of the necessity of Theorem 5.1), that $H_\lambda^n \in (c_0, c)$. Hence, by Exercise 1.3, conditions (5.14) and (8.19) hold.

Sufficiency. Let all of the conditions of the present proposition be satisfied. Then, conditions (5.14) and (8.19) imply, from Exercise 1.3, that $H^n \in (c_0, c)$. Consequently, from (9.11), we can conclude, by (5.15) and (9.10), that the transformation $y = Mx$ exists for every $x \in c_A^\lambda$. $\quad\square$

Theorem 9.1 Let $A = (a_{nk})$ be a λ-reversible method, $B = (b_{nk})$ a triangular method, and $M = (m_{nk})$ an arbitrary matrix. Then, $M \in (c_A^\lambda, c_B^\mu)$ if and only if conditions (5.14), (5.15), (8.19), (9.10), (8.22), and (8.23) are satisfied and

$$\eta, \varphi \in c_G^\mu, \tag{9.12}$$

$$e^k \in c_\Gamma^\mu; \quad \Gamma := (\gamma_{nk}). \tag{9.13}$$

Also, condition (8.22) can be replaced by condition (8.25).

Proof: **Necessity.** Assume that $M \in (c_A^\lambda, c_B^\mu)$. Then, conditions (5.14), (5.15), (8.19), and (9.10) are satisfied by Proposition 9.5, and equality (5.27) holds for every $x := (\xi_k) \in c_A^\lambda$, where $y = (y_k) = (M_k x)$, since B is triangular. This implies that $c_A^\lambda \subset c_G^\mu$. Hence, condition (9.12) is satisfied because $\eta, \varphi \in c_A^\lambda$. As every element ξ_k of a sequence $x := (\xi_k) \in c_A^\lambda$ may be presented in the form (9.8), we can write

$$\sum_{k=0}^{j} g_{nk}\xi_k = \phi \sum_{k=0}^{j} g_{nk}\eta_k + \beta \sum_{k=0}^{j} g_{nk}\varphi_k + \sum_l \frac{\gamma_{nl}^j}{\lambda_l}(\beta_l - \beta) \tag{9.14}$$

for every $x := (\xi_k) \in c_A^\lambda$. From (9.14) it follows, by (9.12), that $\Gamma_\lambda^n := (\gamma_{nl}^j/\lambda_l) \in (c_0, c)$ for every n because A is λ-reversible (see a proof of the necessity of Theorem 5.1). Moreover, from conditions (5.14), (5.15), (8.19), and (9.10), we can conclude that the series $G_n x$ are convergent for every $x \in c_A^\lambda$. Therefore, from (9.14), we obtain

$$G_n x = \phi G_n \eta + \beta G_n \varphi + \sum_l \frac{\gamma_{nl}}{\lambda_l}(\beta_l - \beta) \tag{9.15}$$

for every $x \in c_A^\lambda$. From (9.15), we see, using (9.12), that $\Gamma_\lambda := (\gamma_{nl}/\lambda_l) \in (c_0, c)$. Consequently, condition (8.22) holds, the finite limits γ_l exist, and

$$\lim_n G_n x = \phi\gamma + \beta\psi + \sum_l \frac{\gamma_l}{\lambda_l}(\beta_l - \beta) \tag{9.16}$$

for every $x \in c_A^\lambda$ due to Exercise 1.3. Therefore, we can write

$$\mu_n(G_n x - \lim_n G_n x) = \phi\mu_n(G_n\eta - \gamma) + \beta\mu_n(G_n\varphi - \psi) + \mu_n \sum_l \frac{\gamma_{nl} - \gamma_l}{\lambda_l}(\beta_l - \beta) \tag{9.17}$$

for every $x \in c_A^\lambda$. Using (9.12), it follows from (9.17) that the matrix $\Gamma_{\lambda,\mu} := (\mu_n(\gamma_{nl} - \gamma_l)/\lambda_l) \in (c_0, c)$. Hence, using Exercise 1.3, we conclude that conditions (8.23) and (9.13) are satisfied.

Finally, it is easy to see that condition (8.25) follows from (8.22) and (9.13).

Sufficiency. We assume that all of the conditions of the present theorem hold. Then, the matrix transformation $y = Mx$ exists for every $x \in c_A^\lambda$ by Proposition 9.5. This implies that relations (5.27) and (9.14) hold for every $x \in c_A^\lambda$ (see the proof of the necessity). Using (8.22) and (9.13), we conclude, using Exercise 1.3, that $\Gamma_\lambda^n \in (c_0, c)$ for every n, one can take the limit under the summation sign in the last summand of (9.14). Then, from (9.14), we obtain, by (9.12), the validity of (9.15) for every $x \in c_A^\lambda$. Conditions (9.12)–(9.14) imply that (9.16) holds for every $x \in c_A^\lambda$, due to Exercise 1.3. Then, clearly, relation (9.16) also holds for every $x \in c_A^\lambda$. Moreover, $\Gamma_{\lambda,\mu} \in (c_0, c)$, due to Exercise 1.3. Therefore, $M \in (c_A^\lambda, c_B^\mu)$ by (9.12). $\square$

Using Exercise 9.2, we get the following example for Theorem 9.1, in the case where A is a normal method.

Example 9.2 Let $A = (a_{nk})$ be a normal method, $B = (b_{nk})$ a triangular method and $M = (m_{nk})$ an arbitrary matrix. We show that $M \in (c_A^\lambda, c_B^\mu)$ if and only if conditions (5.14), (8.18), (8.19), (8.22), (8.23), and (9.13) are satisfied and

$$(\rho_n) \in c^\mu, \tag{9.18}$$

$$(\gamma_\lambda^n) \in c^\mu, \tag{9.19}$$

where

$$\rho_n := \lim_j \sum_{l=0}^{j} \gamma_{nl}^j, \quad \gamma_\lambda^n := \lim_j \sum_{l=0}^{j} \frac{\gamma_{nl}^j}{\lambda_l}.$$

In the present case, the methods $H^n := (h_{jl}^n)$ and $\Gamma^n := (\gamma_{nl}^j)$ are lower triangular with

$$h_{jl}^n = \sum_{k=l}^{j} m_{nk} \eta_{kl} \text{ and } \gamma_{nl}^j = \sum_{k=l}^{j} g_{nk} \eta_{kl}, l \le j.$$

Using Exercise 9.2 and (5.7), we obtain

$$\sum_{l=0}^{j} m_{nl} \eta_l = \sum_{l=0}^{j} h_{jl}^n, \sum_{l=0}^{j} m_{nl} \varphi_l = \sum_{l=0}^{j} \frac{h_{jl}^n}{\lambda_l},$$

$$\sum_{l=0}^{j} g_{nl} \eta_l = \sum_{l=0}^{j} \gamma_{nl}^j, \sum_{l=0}^{j} g_{nl} \varphi_l = \sum_{l=0}^{j} \frac{\gamma_{nl}^j}{\lambda_l}.$$

Moreover, it follows from (8.18), that

$$\lim_j \sum_{l=0}^{j} \frac{h_{jl}^n}{\lambda_l}$$

exists and is finite, by the well-known theorem of Dedekind, since the sequence λ^{-1} is monotonically decreasing and bounded. Hence, in the present case, condition (8.18) is equivalent to (5.15) and (9.10). In addition, condition (9.18) is equivalent to (9.12). Thus, $M \in (c_A^\lambda, c_B^\mu)$ by Theorem 9.1.

Remark 9.2 If M is triangular, then conditions (5.14), (5.15), (8.18), (8.19), and (9.10) are redundant in Theorems 9.1 and in Example 9.2.

Definition 9.2 A method A is called λ-perfect, if A is λ-conservative and the set $\{e, e^k, \lambda^{-1}\}$ is fundamental in c_A^λ.

Definition 9.3   A sequence space X is called an FK-space, if X is an F-space (i.e., a complete space with countable system of half-norms separating points in X), where coordinate-wise convergence holds.

We shall need the following result (see [5], p. 57 and [9, 14]).

Lemma 9.1   The domain c_A^λ is an FK-space for an arbitrary method A.

For more information on λ-perfect methods and FK-spaces, the reader can consult, for example, [4–6, 8–9, 11, 13, 14]. We now prove the following result.

Theorem 9.2   Let $A = (a_{nk})$ be a λ-perfect method, $B = (b_{nk})$ a triangular method and $M = (m_{nk})$ an arbitrary matrix. Then, $M \in (c_A^\lambda, c_B^\mu)$ if and only if

$$c_A^\lambda \in m_G^\mu, \tag{9.20}$$

$$e^k, e, \lambda^{-1} \in c_G^\mu. \tag{9.21}$$

Proof: **Necessity.** Assume that $M \in (c_A^\lambda, c_B^\mu)$. Then, clearly, $M \in (c_A^\lambda, m_B^\mu)$, and relation (5.27) holds for every $x \in c_A^\lambda$ since B is triangular. Hence, conditions (9.20) and (9.21) are satisfied since $e^k, e, \lambda^{-1} \in c_A^\lambda$.

Sufficiency. Assume that conditions (9.20) and (9.21) hold. Then, f_n, defined by

$$f_n(x) := \mu_n(G_n x - \lim_n G_n x),$$

is a continuous and linear functional on c_A^λ. In addition, the sequence (f_n) is bounded for every $x \in c_A^\lambda$. In addition, c_A^λ and c_B^λ are FK-spaces. Then (see [13], p. 2 or [11], Corollary 4.22), the set

$$S := \{x \in c_A^\lambda \ : \ \text{there exists the finite limit } \lim_n f_n(x)\}$$

is closed in c_A^λ. Also,

$$\text{lin}\{e^k, e, \lambda^{-1}\} \subset L$$

by condition (9.21), and, due to the λ-perfectness of A, we have

$$c_A^\lambda = cl(\text{lin}\{e^k, e, \lambda^{-1}\}).$$

Consequently, $c_A^\lambda \subset clL = L \subset c_G^\mu$. Thus, $M \in (c_A^\lambda, c_B^\mu)$. $\square$

Corollary 9.2   If A is a normal method and B and M are triangular matrices, then condition (9.20) in Theorem 9.2 can be replaced by the condition $c_A^\lambda \in m_G^\mu$.

Proof: With $\Gamma := GA^{-1}$, we can write $B_n y = G_n x = \Gamma_n z$ for every $x \in c_A^\lambda$, where the transformation $y = Mx$ exists and $z = Ax \in m^\lambda$. For each $z \in m^\lambda$ $(z \in c^\lambda)$

there exists an $x \in m_A^\lambda$ ($x \in c_A^\lambda$, respectively), such that $z = Ax$ since the normal method A is also reversible, $c^\lambda \subset m^\lambda \subset c$ and $c_A^\lambda \subset m_A^\lambda \subset c_A$. This implies that $M \in (c_A^\lambda, m_B^\mu)$ is equivalent to $c^\lambda \subset m_\Gamma^\mu$, and $M \in (m_A^\lambda, m_B^\mu)$ is equivalent to $m^\lambda \subset m_\Gamma^\mu$. From Exercise 8.3, we conclude that $c^\lambda \subset m_\Gamma^\mu$ if and only if $m^\lambda \subset m_\Gamma^\mu$.

For $A = (R, p_n)$, we obtain the following example.

Example 9.3 Let $B = (b_{nk})$ be a triangular method, $M = (m_{nk})$ an arbitrary matrix, and (R, p_n) the conservative method, satisfying property (8.36). We shall prove that $M \in (c_{(R,p_n)}, c_B^\mu)$ if and only if conditions (7.10), (7.11), and (7.83) are satisfied and

$$e^k, e, \lambda^{-1} \in c_G^\mu, \tag{9.22}$$

$$\mu_n \sum_l \left| P_l \Delta_l \frac{\Delta_l(g_{nl} - g_l)}{p_l} \right| = O(1). \tag{9.23}$$

Assume that $M \in (c_{(R,p_n)}, c_B^\mu)$. Using (7.9), we can check that relations (7.13), (7.14), and (7.84) are satisfied, and

$$\gamma_{nl}^j = \begin{cases} \gamma_{nl} & (l < j\text{-}1), \\ \gamma_{n,j-1} - P_j g_{n,j+1}/p_j + g_{n,j+1} & (l = j\text{-}1), \\ P_j g_{nj}/p_j & (l = j), \\ 0 & (l > j) \end{cases} \tag{9.24}$$

hold. Hence, using Example 9.2, we obtain that conditions (7.10), (7.11), (7.83), (9.22), and (9.23) are satisfied (in the present case, we can take $\lambda_l = 1$). Conversely, let conditions (7.10), (7.11), (7.83), (9.22), and (9.23) be satisfied. We shall show that all of the conditions presented in Example 9.2 are satisfied. Using (7.84), we conclude that conditions (8.22), (8.23), and (9.13) are satisfied by (7.83), (9.22), and (9.23). Also conditions (5.14), (8.18), and (9.18) hold (see Hint of Exercise 8.6). Moreover, condition (9.19) coincides with (9.18) since $\lambda_l \equiv 1$. To show that condition (8.19) also holds, we, using (7.13), can write

$$\sum_l |h_{jl}^n| \leq \sum_{l=0}^{j-1} |h_{nl}| + \left| \frac{P_j m_{n,j+1}}{p_j} \right| + |m_{n,j+1}| + \left| \frac{P_j m_{nj}}{p_j} \right|. \tag{9.25}$$

Since (R, p_n) is conservative, condition (7.15) holds (see the proof of Proposition 7.1). In addition,

$$\frac{P_j m_{n,j+1}}{p_j} = O_n(1) \frac{P_{j+1} m_{n,j+1}}{p_{j+1}} = O_n(1)$$

by (8.36) and (7.11). Therefore, from (9.25) (using (7.10) and (7.11)), we obtain that condition (8.19) also holds. Thus, $M \in (c_{(R,p_n)}, c_B^\mu)$, due to Example 9.2.

We now describe a class of matrices, consisting of lower triangular matrices M defined by equation (7.82), which transform c_A^λ into c_B^λ for $A = C^\alpha$ and $B = C^\beta$.

Example 9.4 Let α, β be numbers satisfying $\alpha > 0, Re\ \beta > 0$, and $M = (m_{nk})$ a lower triangular matrix defined by (7.82). Let $\lambda := (\lambda_k)$ and $\mu := (\mu_k)$ be sequences defined by $\lambda_k := A_k^\alpha$ and $\mu_k := (k+1)^s$ with $s > 0$. We prove that if

$$s \le Re\ \beta; \text{ and } Re\ r < -s - 1,$$

then $M \in (c_{C^\alpha}^\lambda, c_{C^\beta}^\mu)$. We shall show that all of the conditions of Example 9.2 are satisfied for $A = C^\alpha$ and $B = C^\beta$. It is easy to see that λ and μ are positive monotonically increasing unbounded sequences, and conditions (5.14), (8.18), and (8.19) are valid by Remark 9.2. Using (7.48), we obtain that $G = (g_{nk})$ is a lower triangular matrix, where

$$g_{nk} = \frac{A_{n-k}^{\beta+r+1}}{A_n^\beta}, k \le n. \tag{9.26}$$

Since $a_{n0} \equiv 1$, then $\rho_n = g_{n0}$ (see Hint of Exercise 8.6). Hence,

$$\rho_n = \frac{A_n^{\beta+r+1}}{A_n^\beta} = O(1)(n+1)^{Re\ r+1}$$

by (7.46) and (7.47). This implies $\lim_n \rho_n = 0$ because $Re\ r + 1 < 0$. Therefore,

$$\mu_n(\rho_n - \lim_n \rho_n) = O(1)(n+1)^{s+Re\ r+1} = o(1)$$

since $s + Re\ r + 1 < 0$, that is, condition (9.18) is fulfilled. Using (7.50), we can write

$$\sum_{l=0}^j \frac{\gamma_{nl}^j}{\lambda_l} = \sum_{k=0}^j g_{nk} \sum_{l=0}^k \frac{A_l^\alpha A_{k-l}^{-\alpha-2}}{A_l^\alpha} = \sum_{k=0}^j g_{nk} A_l^{-\alpha-1}.$$

Then,

$$\gamma_\lambda^n = \sum_{k=0}^n g_{nk} A_l^{-\alpha-1} = \frac{A_n^{\beta-\alpha+r+1}}{A_n^\beta}$$

by (9.26) and (7.48). As

$$\gamma_\lambda^n = O(1)(n+1)^{-\alpha+Re\ r+1}$$

by (7.46) and (7.47), and $-\alpha + Re\ r + 1 < 0$, then $\lim_n \gamma_\lambda^n = 0$. Consequently, due to $s - \alpha + Re\ r + 1 < 0$ we have

$$\mu_n(\gamma_\lambda^n - \lim_n \gamma_\lambda^n) = \mu_n \gamma_\lambda^n = O(1)(n+1)^{s-\alpha+Re\ r+1} = o(1),$$

that is, condition (9.19) is satisfied.

Using (7.43), (7.48), and (9.26), we obtain that $\Gamma = (\gamma_{nk})$ is a lower triangular matrix, where

$$\gamma_{nl} = \frac{A_l^\alpha A_{n-l}^{\beta+r-\alpha}}{A_n^\beta}, l \leq n. \tag{9.27}$$

This implies, by (7.46) and (7.47), that

$$\gamma_{nl} = O(1)(l+1)^\alpha \left(\frac{n-l+1}{n+1}\right)^{Re\ \beta} (n-l+1)^{Re\ r-\alpha} = O_l(1)(n-l+1)^{Re\ r-\alpha}.$$

Since $Re\ r - \alpha < 0$, we get $\gamma_k = 0$. Consequently,

$$\mu_n(\gamma_{nl} - \gamma_l) = O_l(1)(n+1)^s(n-l+1)^{Re\ r-\alpha} = O_l(1)(n+1)^{s+Re\ r-\alpha} = o_l(1),$$

since $s + Re\ r - \alpha < 0$, that is, condition (9.13) is fulfilled. Using (9.27) we obtain

$$T_n := \mu_n \sum_{l=0}^n \frac{|\gamma_{nl}|}{\lambda_l} = \frac{(n+1)^s}{|A_n^\beta|} \sum_{l=0}^n |A_{n-l}^{\beta+r-\alpha}|.$$

Therefore,

$$T_n = O(1)(n+1)^{s-Re\ \beta} \sum_{l=0}^n (n-l+1)^{Re\ (\beta+r)-\alpha}$$

$$= O(1)(n+1)^{s-Re\ \beta} \sum_{l=0}^n (l+1)^{Re\ (\beta+r)-\alpha}$$

by (7.46) and (7.47). Further, for $s < Re\ \beta$, we consider three different cases.

1) For $Re\ (\beta+r) - \alpha < -1$, we have $T_n = o(1)$.
2) For $Re\ (\beta+r) - \alpha = -1$, we obtain

$$T_n = O(1)(n+1)^{s-Re\ \beta} \ln\ (n+1) = o(1).$$

3) For $Re\ (\beta+r) - \alpha > -1$, we obtain

$$T_n = O(1)(n+1)^{s+Re\ r-\alpha+1} = o(1)$$

since $s + Re\ r - \alpha + 1$.

For $s = Re\ \beta$, we have $T_n = O(1)$ due to $Re\ (\beta+r) - \alpha < -1$. So we can conclude that conditions (8.22) and (8.23) also hold (recall that $\gamma_k = 0$). Thus, $M \in (c_{C^\alpha}^\lambda, c_{C^\beta}^\mu)$, due to Example 9.2.

Theorems 9.1 and 9.2 and Corollaries 9.2 and 9.3 were first proved in [1]. For summability factors from $c_{(R,p_n)}^\lambda$ into c_B^μ, necessary and sufficient conditions are found in [7].

9.4 Excercise

Exercise 9.1   Which form does (9.8) take if A is a sequence-to-sequence Cesàro method, that is, $A = \tilde{C}^\alpha$?
Hint. See Remark 9.1.

Exercise 9.2   Prove that, if a method A is normal, then

$$\varphi_k = \sum_{l=0}^{k} \frac{\eta_{kl}}{\lambda_l},$$

where $\varphi := (\varphi_k)$ and $\eta := (\eta_{kj})$ for each fixed j are the solutions of the system $y = Ax$, respectively, for $y = (\delta_{nn}/\lambda_n)$ and $y = (y_n) = (\delta_{nj})$, and $(\eta_{kn}/\lambda_n) \in l$ for every fixed k.
Hint. For a normal method A, relation (9.8), we can present in the form

$$\xi_k = \left(\eta_k - \sum_{l=0}^{k} \eta_{kl} \right) \phi + \left(\varphi_k - \sum_{l=0}^{k} \frac{\eta_{kl}}{\lambda_l} \right) \beta + \sum_{l=0}^{k} \eta_{kl} A_l x$$

for every $x \in c_A^\lambda$, where, in the present case, (η_{kl}) is the inverse matrix of A. Use Proposition 5.4.

Exercise 9.3   Let $A = (a_{nk})$ be a λ-reversible method, $B = (b_{nk})$ a triangular method, and $M = (m_{nk})$ an arbitrary matrix. Prove that $M \in (c_A^\lambda, m_B^\mu)$ if and only if conditions (5.14), (5.15), (5.25), (8.19), (8.22), (8.23), and (9.10) are satisfied and $\eta, \varphi \in m_G^\mu$.
Hint. Use the relation $c_B^\mu \subset m_B^\mu$) and the proof of Theorem 9.1.

Exercise 9.4   Let $A = (a_{nk})$ be a normal method, $B = (b_{nk})$ a triangular matrix, and $M = (m_{nk})$ an arbitrary matrix. Prove that $M \in (c_A^\lambda, m_B^\mu)$ if and only if conditions (5.14), (5.25), (8.18), (8.19), (8.22), and (8.23) are satisfied and $(\gamma_\lambda^n) \in m^\mu$,

$$(\rho_n) \in m^\mu. \tag{9.28}$$

Hint. Use Exercise 9.3. See also Example 9.2. We note also that the existence of the finite limits γ_λ^n follows from the existence of the finite limits ρ_n, by Dedekind's theorem.

Exercise 9.5   Prove that, if a method B is normal, then condition (8.19) is redundant in Theorem 9.1, Example 9.2, and in Exercises 9.3 and 9.4.

Exercise 9.6   Prove that, if a method $A = (a_{nk})$ has the property $a_{n0} \equiv 1$, then condition (8.18) is redundant in Exercises 9.4 and 9.5, and condition (9.18) (condition (9.28)) can be replaced by condition $e^0 \in c_G^\mu$ (by condition (8.37), respectively).

Exercise 9.7   Prove that, if B is a normal method, then condition (7.10) is redundant in Example 9.3.

Exercise 9.8   Let α, β be complex numbers satisfying the properties $\alpha \neq -1, -2, \ldots, Re\, \beta > 0$, and $M = (m_{nk})$ a lower triangular matrix defined by (7.82). Let $\mu := (\mu_k)$ be a sequence defined by $\mu_k := (k+1)^s$ with $0 < s < Re\, (\beta - \alpha) - 1$. Prove that, if

$$s \leq Re\, \beta \text{ and } Re\, r \leq Re\, (\alpha - \beta),$$

then $M \in (c_{C^\alpha}, c_{C^\beta}^\mu)$.

Hint. Show that all of the conditions presented in Example 9.2 are satisfied, for $A = C^\alpha, B = C^\beta$ and $\lambda_k \equiv 1$.

References

1 Aasma, A.: Matrix transformations of λ-summability fields of λ-reversible and λ-perfect methods. Comment. Math. Prace Mat. **38**, 1–20 (1998).

2 Beekmann, W. and Chang, S.-C.: λ-convergence and λ-conullity. Z. Anal. Anwend. **12(1)**, 179–182 (1993).

3 Beekmann, W. and Chang, S.-C.: λ-convergence and λ-replaceability. Tamkang J. Math. **26(2)**, 145–147 (1995).

4 Boos, J.: Classical and Modern Methods in Summability. Oxford University Press, Oxford (2000).

5 Jürimae, E.: Properties of matrix mappings on rate-spaces and spaces with speed. Tartu Ül. Toimetised **970**, 53–64 (1994).

6 Jürimäe, E.: Matrix mappings between rate-spaces and spaces with speed. Tartu Ül. Toimetised **970**, 29–52 (1994).

7 Kangro, G.: O množitelyah summirujemosti tipa Bora-Hardy dlya zadannoi skorosti I (On the summability factors of the Bohr-Hardy type for a given speed I). Eesti NSV Tead. Akad. Toimetised Füüs.-Mat. **18(2)**, 137–146 (1969).

8 Kangro, G.: O λ-sovershennosti metodov summirovanya i ee primenenyah. II. (The λ-perfectness of summability methods and applications of it. II.). Eesti NSV Tead. Akad. Toimetised Füüs.-Mat. **20**, 375–385 (1971).

9 Kangro, G.: O λ-sovershennosti metodov summirovanya i ee primenenyah. I. (The λ-perfectness of summability methods and applications of it. I.). Eesti NSV Tead. Akad. Toimetised Füüs.-Mat. **20**, 111–120 (1971).

10 Kangro, G.: Množiteli summirujemosti dlya ryadov, λ-ograniťšennõh metodami Rica i Cezaro (Summability factors for the series λ-bounded by the methods of Riesz and Cesàro). Tartu Riikl. Ül. Toimetised **277**, 136–154 (1971).

11 Leiger, T.: Funktsionaalanalüüsi meetodid summeeruvusteoorias (Methods of functional analysis in summability theory). Tartu Ülikool, Tartu (1992).

12 Leiger, T. and Maasik, M.: O λ-vklyutchenyy matriz summirovaniya (The λ-inclusion of summation matrices). Tartu Riikl. Ül. Toimetised **770**, 61–68 (1987).

13 Wilansky, A. Summability through Functional Analysis, North-Holland Mathematics Studies, Vol. 85; Notas de Matemática (Mathematical Notes), Vol. 91. North-Holland Publishing Co., Amsterdam (1984).

14 Zeller, K.: Allgemeine Eigenschaften von Limitierungsverfahren. Math. Z. **53**, 463–487 (1951).

Index

An Introductory Course in Summability Theory, First Edition. Ants Aasma, Hemen Dutta, and P.N. Natarajan.
© 2017 John Wiley & Sons, Inc. Published 2017 by John Wiley & Sons, Inc.